Of Maybugs and Men

Of Maybugs and Men

A HISTORY AND PHILOSOPHY OF
THE SCIENCES OF HOMOSEXUALITY

Pieter R. Adriaens and
Andreas De Block

THE UNIVERSITY OF CHICAGO PRESS
CHICAGO AND LONDON

The University of Chicago Press, Chicago 60637
The University of Chicago Press, Ltd., London
© 2022 by The University of Chicago
All rights reserved. No part of this book may be used or reproduced in any manner whatsoever without written permission, except in the case of brief quotations in critical articles and reviews. For more information, contact the University of Chicago Press, 1427 E. 60th St., Chicago, IL 60637.
Published 2022
Printed and bound by CPI Group (UK) Ltd, Croydon, CR0 4YY

31 30 29 28 27 26 25 24 23 22   1 2 3 4 5

ISBN-13: 978-0-226-82242-6 (cloth)
ISBN-13: 978-0-226-82244-0 (paper)
ISBN-13: 978-0-226-82243-3 (e-book)
DOI: https://doi.org/10.7208/chicago/9780226822433.001.0001

Published with the support of the Belgian University Foundation.

Library of Congress Cataloging-in-Publication Data

Names: Adriaens, Pieter R., author. | Block, Andreas de, author.
Title: Of maybugs and men : a history and philosophy of the sciences of homosexuality / Pieter R. Adriaens and Andreas De Block.
Other titles: History and philosophy of the sciences of homosexuality
Description: Chicago ; London : The University of Chicago Press, 2022. | Includes bibliographical references and index.
Identifiers: LCCN 2022022554 | ISBN 9780226822426 (cloth) | ISBN 9780226822440 (paperback) | ISBN 9780226822433 (ebook)
Subjects: LCSH: Male homosexuality—Research—History. | Male homosexuality—Research—Philosophy. | Homosexuality—Genetic aspects—Research—History. | BISAC: SOCIAL SCIENCE / LGBTQ Studies / Gay Studies | SCIENCE / Life Sciences / Biology
Classification: LCC HQ76 .A37 2022 | DDC 306.76/62—dc23/
eng/20220602
LC record available at https://lccn.loc.gov/2022022554

♾ This paper meets the requirements of ANSI/NISO Z39.48-1992 (Permanence of Paper).

For Joris and Joris, with whom it probably all began. (PRA)

Contents

Introduction: Thinking about
Science and Homosexuality   *1*

CHAPTER 1   Not by Genes and Hormones Alone:
On Homosexuality and Innateness   *16*

CHAPTER 2   Sham Matings and Other Shenanigans:
On Animal Homosexuality   *50*

CHAPTER 3   Beyond the Paradox: On Homosexuality
and Evolutionary Theory   *99*

CHAPTER 4   Values, Facts, and Disorders:
On Homosexuality and Psychiatry   *136*

Epilogue: Gaydars and the Dangers of
Research on Sexual Orientation   *175*

Acknowledgments   193

Notes   195

References   205

Index   233

Introduction
Thinking about Science and Homosexuality

A knowledge of the historic and philosophical background gives that kind of independence from prejudices of his generation from which most scientists are suffering.

ALBERT EINSTEIN

"Why do some men enjoy submitting to sex?" The question would not be out of place in a contemporary science magazine ("What makes people gay?"), but it in fact originates with an unknown Greek philosopher from late antiquity, in a treatise (the *Problemata*) inspired by the Aristotelian tradition of natural philosophy.[1] (Readers will find answers to this question at the beginning of chapter 1.) The core of this tradition was to describe and explain the wide variety of natural phenomena, and for natural philosophers that variety certainly included human male homosexuality.

The ancient Greeks may have been quite casual about some homosexual activities (at least if their vases are anything to go by), but between the eleventh and the thirteenth century, when many Greek and Arabic texts, including the *Problemata*, were translated into Latin, sexual morality had changed dramatically. Sexual acts between men were now subsumed under the sin of sodomy and denounced by ecclesiastical and worldly authorities alike. One can only imagine, then, the reactions of late medieval readers of the *Problemata*, many of whom were loyal fans of Aristotle, when confronted with this time capsule's frank questions about anal sex between men.

Thanks to historians of philosophy, we actually do know what some of them were thinking (Cadden 2001). The fourteenth-century Italian philosopher and commentator Pietro d'Abano was clearly in two minds about the *Problemata*'s discussion. On the one hand, he dauntlessly continued its naturalistic search for causes of male homosexuality, adding his own astrological explanations along the way. On the other, and slightly at odds with the overall evenhandedness of his comments, he appended a new and rather moralistic narrative, condemning male homosexuality as a "perverse and filthy habit" and abhorring the "monstrous nature" of homosexuals (quoted in Cadden 2001, 75–76).

D'Abano's work can itself serve as a time capsule for contemporary commentators who, on the face of it, will perhaps find few affinities with his late medieval mindset. Over time, authority-based astrological hypotheses have given way to neurological, endocrinological, and genetic hypotheses. Moreover, d'Abano's crude condemnation of homosexuality would seem to set him light-years apart from contemporary scientific research on the topic, which prides itself in pursuing value-free knowledge. In reality, however, the ideal of value-free science has often merely pushed moral values underground, where their effects are much harder to assess. The psychiatric history of homosexuality—the central topic of the final chapter—is a case in point. Scientific evidence may have played a role in both establishing and abandoning the view that homosexuality is a mental disorder, but so did a number of value judgments made by psychiatrists, including the judgment that heterosexuality is the sexual norm.

This book is all about the intimate engagement of facts and values in our thinking about homosexuality during the past two hundred years. More particularly, it focuses on the body of knowledge provided by a range of "gay sciences" scattered across various science groups, including the humanities, the social sciences, and the biomedical sciences.[2] Still, this is *not* a science book. Rather, it is a book about the history and philosophy of the gay sciences. In this introduction, we elucidate the nature and the inevitable limitations of our approach, we discuss our terminology, and we provide a brief summary of the book's four chapters and its epilogue.

HISTORY AND PHILOSOPHY OF THE GAY SCIENCES

Science produces reliable knowledge, and so twenty-first-century Westerners turn to the sciences to discover more about the nature and causes of homosexuality. We appeal to life scientists in our search for the biological controls of homosexuality; we call in psychologists to assess the well-being or social success of children raised by gay parents; we rely on historians and anthropologists to map out the sexual landscapes of earlier and non-Western societies. We trust their answers more than those provided by any other source of knowledge, including folk science, theology, or gay magazines.

Still, the gay sciences are many, and dissensus has been legion, both in the past and in the present, and both within and among disciplines. This book details a number of such disputes, and it does so by meandering between two disciplinary domains: the history of science and the philosophy of science. In this section, we consider the value and functions of both domains in turn.

Historians and philosophers of science often consider the history of science a "graveyard of science" (Frost-Arnold 2011, 1138) or a "cemetery of theories" (Stengers 2000, 31) that may have worked for a while but were then abandoned for a better alternative. Why should anyone be interested in such a cemetery? However, we agree with historians of science that their work can generally be useful in many ways (see, e.g., Chang 2017), some of which we hope to illustrate in this book.

First, a history of science perspective can help excite curiosity about the process of past scientific endeavors. How did science proceed in the past? What kinds of evidence were being considered in scientific decision-making? One wonders, for example, how psychiatrists came to consider homosexuality as a disorder rather than as a sin or a crime, which had previously been common practice. What arguments, if any, did they present to make their case? And how did they come to abandon that view again in the early 1970s?

Second, studying the history of science can improve our understanding of current scientific knowledge and practice by carefully reconstructing its historical trajectory. How did scientists come to think this way? For example, knowing how and why twentieth-century psychiatrists abandoned the disorder view of homosexuality may help us understand how their contemporary colleagues think about (historically) related conditions such as fetishism and pedophilia and why they continue to distinguish between disordered and nondisordered sexual interests.

Third, the history of science can actually improve contemporary science itself, rather than just our understanding of it, and function as what philosopher Hasok Chang (2017, 94) calls "a complementary science." In that capacity, it can confront scientists with the past contingencies behind their orthodoxy,[3] expose their blind spots and their indebtedness to what is now considered flawed science, and fill the gaps in their work. The success and focus of many contemporary scientists stems from their ruthless elimination of alternative questions and perspectives, some of which are perfectly scientific in themselves. For example, many contemporary biologists prefer mechanistic to mentalistic explanations of animal homosexual behaviors. They find it hard to even consider attributing these behaviors to underlying mental states, such as desires or preferences, and opt instead for explanations in terms of inabilities, unusual circumstances, and overwhelming biological imperatives. In chapter 2, we will see that the work of some nineteenth-century zoologists can help us break that trend. More generally, several chapters in this book can be read as an appeal to at least consider alternative questions and hypotheses in the gay sciences, even if they seem unconventional, outdated, or downright incongruous.

Questions about usefulness have been asked not only about the history of science but also about the philosophy of science—a second domain in which this book operates. Some scientists are very dismissive of philosophers who study their discipline. The Nobel Prize–winning physicist Richard Feynman was among them. He once allegedly quipped that the philosophy of science is as useful to scientists as ornithology is to birds.[4] Chang (2017, 91) adds, "I fear what he might have thought about the history of science, which would be more like paleontology, obsessing about extinct birds." A crash course in ornithology might indeed be very helpful to some birds, if only their brains would allow it, but that, of course, is not what Feynman meant. Rather, he meant to say that philosophers (of science) are of no use to scientists.

We disagree with Feynman, and we are not the only ones. In a recent opinion paper by Lucie Laplane et al. (2019), some philosophers and scientists teamed up to advocate the importance of the philosophy of science for the sciences. In their view, its contribution can take at least three forms.

First, philosophers have always been concerned with analyzing, clarifying, and even improving the concepts we use in various spheres of life. Some philosophers even claim that such conceptual work is "the proper task of . . . philosophy" (P. Strawson 1985, 23). Of course, there are scientists who are happy to perform their own conceptual work, but very often their diligence prevents them from indulging in it, and that is where philosophers of science can step in. Likewise, many chapters of this book aim to analyze and clarify the concepts used in the gay sciences, including the ideas of innateness, desire, preference, and disorder.

Second, philosophers of science critically engage with the assumptions that underlie scientific hypotheses and explanations. Much like their concepts, over time, many of the assumptions scientists build on have become so commonplace as to be virtually invisible. Philosophers of science can then alert us when these assumptions are or become problematic in the light of new scientific or philosophical developments. If it is true, for example, that many animals are capable of having mental states formerly assumed to be uniquely human, as recent philosophical work on animal minds suggests (see, e.g., Andrews 2014), then it might be time to reconsider the assumption that mechanistic explanations of homosexual behavior in animals are always superior to mentalistic explanations.

Third, philosophers of science can help foster dialogue among the sciences, another pursuit for which many scientists have neither time nor interest. Academic research today encourages specialization, and specialization can come at the expense of interdisciplinarity. In the third chapter of this book, for example, we argue that some contemporary evolutionists overestimate the importance of the particular form of homosexuality with which twenty-first-century Westerners are most familiar. In doing so, they neglect or disregard other forms of homosexuality that are extensively documented by historians and anthropologists. By bringing together insights from various disciplines in both the life sciences and the human sciences, we hope to set up new ways to research the evolution of homosexuality.[5]

The dialogue stimulated by philosophers of science benefits not only the sciences but also society. Science does not occur in a vacuum. It is as much the product of a society as it produces and molds that society. Very often, its findings impact the way we think about even the most intimate parts of our lives, including our sexuality. Traditionally, scientific findings about homosexuality have figured prominently in moral and political debates. Some years ago, for example, the author of a catalogue for a prominent exhibition on animal homosexuality noted that "homosexuality is a common and widespread phenomenon in the animal world," and he expressed his hope that this finding would finally "reject the all too well-known argument that homosexual behavior is a crime against nature" (Søli 2009). Discussing recent work in genetics, evolutionary biologist Emmanuele A. Jannini and his colleagues (2010, 3250) offered a similar argument: "These [genetic] studies suggest that the dynamics of genetic factors influencing homosexuality . . . are a natural aspect of human sexual variability. These findings further discredit the assumptions that homosexuality is pathological and that it should be cured rather than accepted and respected." Such arguments are perhaps as problematic as they are understandable. They are problematic because of their basic assumption that nature can provide us with moral guidelines—an assumption we will return to in the epilogue of this book. Still, they are understandable because scientists and science educators are human beings, and their ideas are to some extent informed by their everyday attitudes and social interactions—attitudes and interactions in which moral and other nonepistemic values play an important role.

As philosophers of science, we cherish the intricate engagement between science and values. Science is not, cannot, and should not be value-free. Its values do sometimes bias our understanding of sexuality, and we are well aware that in the past, such biases have caused a lot of misery, particularly for homosexuals. Sometimes, however, discussing values can help us get a better picture of how the world really is. Some work in feminist philosophy of science, for example, builds on the idea that the patriarchy has produced a lot of bad science, but it also acknowledges that the mistakes of the past can be corrected by incorporating a feminine perspective into science (Lloyd 2005). Similarly, some of the current sciences of sexual orientation may be more reliable because they generally build on a less negative attitude toward homosexuality than the nineteenth- or twentieth-century sciences of sexual orientation.

All in all, we believe that both the history of science and the philosophy of science can be very useful for the gay sciences. Some readers may wonder, however, why we would want to combine both disciplines. The answer is that they are complementary sciences, both in Hasok Chang's sense that they can complement the sciences and in the sense that they can complement each other. As mentioned before, the history of science can be quite useful to various actors, and that includes philosophers of science. Only the history of science can really substantiate, corroborate, or falsify philosophical claims about the progress, utility, and the rules of the scientific game. The famous philosopher of science Imre Lakatos (1970, 91) even went so far as to say that "philosophy of science without history of science is empty." Conversely, according to Lakatos, "history of science without philosophy of science is blind." We always look at history from some theoretical perspective; the history of science can only increase our knowledge when we are able to recognize the relevance of historical facts.

A LOVE THAT SPEAKS ITS MANY NAMES

The Austro-Hungarian journalist and human rights activist Károly Kertbeny coined the word "homosexuality" (*Homosexualität*) in an 1868 private letter to his German colleague, Karl Heinrich Ulrichs, whom we will meet again later in several chapters of this book (Féray, Herzer, and Peppel 1990). Both Kertbeny and Ulrichs were active anonymously in denouncing the criminal prosecution of homosexuals under Prussian and German law, although they did so on different grounds. While Kertbeny opted for the liberal argument that the state does not have the right to meddle in sexual relationships between consenting adult partners (Takács 2004), Ulrichs advocated the naturalistic argument that homosexuality's innateness should exempt it from punishment.

Kertbeny's neologism "homosexuality" was probably connected to his activist struggle (Herzer 1986). While he never explicitly provided a reason for introducing it, his writing suggests that he considered some of the existing terminology either stigmatizing, as in the case of "sodomy," "pederasty," and "indecency against nature" (*widernatürliche Unzucht*), or hopelessly confused, as in the case of "sexual inversion," "contrary sexual feeling" (*konträre Sexualempfindung*), and Ulrichs's "Uranism." The long-term success of Kertbeny's term was likely due to its generality and its theory neutrality. Because "homosexuality" did not spring from any theory about sexuality, it readily found its place in the work of various (and even rivaling) early twentieth-century authors (Herzer 1986, 16). Moreover, following Kertbeny's own practice, many of them quickly interpreted the term in the broadest possible sense. In his *Studies in the Psychology of Sex*, for example, sexologist Henry Havelock Ellis proposed to distinguish homosexuality from its various subtypes, such as (the innate) sexual inversion: "Sexual inversion . . . means sexual instinct turned by inborn constitutional abnormality toward persons of the same sex. It is thus a narrower term than homosexuality, which includes all sexual attractions between persons of the same sex, even when seemingly due to the accidental absence of the natural objects of sexual attraction, a phenomenon of wide occurrence among all human races" (H. Ellis 1900, 11).

In this book, we follow Kertbeny's neutral and broad interpretation of the term "homosexuality" as referring to any kind of sexual state or activity, whether physical or mental, that involves or focuses on one or more persons of the same sex. One of the advantages of this thin definition is that it allows us to ascribe homosexuality to a variety of states, activities, and actors and to talk about various forms of homosexuality. When it comes to actors, our definition allows us to include nonhuman animals and individuals from non-Western or past societies, even if some of their activities markedly differ from those we typically associate with contemporary Western homosexuality. When it comes to states and activities, our definition allows us to meaningfully talk about homosexual behaviors as well as about homosexual desires, fantasies, preferences, orientations, and even identities. In chapter 2, we go to great lengths in defining and demarcating these various states and activities, examining some rather unusual combinations of actors and activities. Some philosophers and biologists firmly believe, for example, that there are no such things as homosexual desires and preferences among nonhuman animals. We will argue against this view on both conceptual and empirical grounds. As it turns out, many nonhuman animals are motivated to have sex with individuals of their own sex and hence can be said to have homosexual desires, and at least some animals prefer homosexual activities, since they choose to have sex with individuals of their own sex even when individuals of the other sex are sexually available. Whether animals can have a homosexual orientation, here defined as a stable homosexual preference, is a hotly debated matter, but we think that the evidence supports the view that such orientation can be found in at least a few species.

Finally, our definition of homosexuality is so thin that it allows us to distinguish among diverse forms of homosexuality in men, or homosexual phenotypes, three of which play quite important roles in this book. Some incarcerated men are known to engage in homosexual behaviors and to fantasize about same-sex partners, even though most of them do not identify as homosexuals, and even though they would sexually prefer female to male partners if they were given a choice. Such prison homosexuality is often described as one example of *situational homosexuality* (Hartenstein and Gonsiorek 2015). Another homosexual phenotype is known as *modern* or *egalitarian homosexuality* (Murray 2000). This is the form of homosexuality with which twenty-first-century Westerners are most familiar, as modern homosexuals are perhaps the most visible and vocal subgroup of all homosexuals. Modern homosexual men have sex with men because they consistently prefer homosexual to heterosexual activities, even to the point of being exclusively homosexual throughout their lives. Moreover, most of these men self-identify as homosexuals, and they often claim to have discovered their sexual orientation at an early age. Interestingly, modern homosexual partners tend to be of similar age and social status (and thus egalitarian), which distinguishes modern homosexuality from a third phenotype. In many societies, both past and present, adult men engage in sexual activities with boys and adolescents. They prefer these activities to heterosexual activities, even though society expects them to marry and have children, and even though they do not self-identify as homosexuals. This third form of homosexuality is known under many names, including *transgenerational homosexuality* (Stein 1999) and *ritualized homosexuality* (Herdt 1994).[6]

Many scientists have no trouble naming these various (combinations of) activities and states as instances of homosexuality. Kertbeny's neologism, however, has also had its opponents, both within the (gay) scientific literature and beyond. LGBTQ studies, for example, tend to shun the term because many believe it suggests a dichotomous or binary difference between homosexuality and heterosexuality—a difference that simply does not fit with the way many people experience and understand their own sexuality (Murphy 1997). Some zoologists have suggested reserving "homosexuality" for human activities and replacing it in the context of animal behavior with supposedly more neutral terms like "isosexuality" or "same-sex sexuality" (Gowaty 1982). Meanwhile, anthropologists have proposed more specific terms to refer to particular non-Western homosexual activities. Anthropologist Gilbert Herdt (1994, xiv), for example, once suggested renaming the Sambia people's ritualized homosexual activities as "boy-inseminating practices." In chapter 2, we will discuss some of these proposals in more detail.

In recent years, this academic chorus has expanded to include some leading American news organizations and newspapers, whose stylebooks recommend substituting "homosexual" with "gay" for reasons not dissimilar to Kertbeny's in the 1860s. In their view, the word "homosexual" should be avoided because it is "derogatory and offensive" (GLAAD 2016, 6). More particularly, it is "a word whose clinical history and pejorative connotations are routinely exploited by anti-LGBTQ extremists to suggest that people attracted to the same sex are somehow diseased or psychologically and emotionally disordered" (15). In short, we should avoid using "homosexuality" because of the negative connotations it acquired in the hands of nineteenth and twentieth-century psychiatrists and anti-gay crusaders, and because it may predispose people to negative attitudes toward homosexuality and homosexuals (B. Smith et al. 2018).

We disagree with these recommendations, for reasons we explore here and in some of the chapters of this book.

First, any term can be used pejoratively or amelioratively. Kertbeny's term has always been in the vocabulary of both advocates and opponents of homosexuality's normality or moral value. Who decides, then, which connotation gets the upper hand? One popular answer to this question is that the decision is up to homosexuals themselves, as people have the right to be called whatever they choose. However, some philosophers have pointed out that this rule perhaps applies to proper names ("Lady Gaga") and quasi-proper names of classes ("African American") but not, or at least not evidently, to descriptive predicates, such as "homosexual." The American philosopher Christopher Boorse (2010, 79) gives a personal example, noting how "philosophy professors have no right to be called 'persons of profundity,' 'moral paragons,' 'natural rulers,' etc., no matter how much they might enjoy these descriptions." Importantly, minority activists have shown that terms can be reclaimed and that it is often easier to shift meanings than to replace words (Barnes 2016, 8). "Queer," for example, was introduced in the English language as a derogatory term but is now embraced by the gay community and associated with pride and self-confidence (Berlant et al. 1994).

Second, changing our vocabulary may alleviate some people's concerns in the short term, but in the long term, it produces a terminological volatility that is undesirable in scientific research. In the past, scientists have built a whole repertory of designations to refer to sexual activities between partners of the same sex, and each new designation increases the risk of obscuring interesting continuities and parallels among findings. We would rather not continue that tradition.

Third, none of the suggested alternative terms strike us as a worthy replacement for "homosexuality." For one thing, they all lack its remarkable versatility. Many languages have some variant of the word "homosexuality," which is not the case for "gay," unless as a loan word. Many languages also have several lexical categories of the word: like "gay," "homosexual" can act as both a noun and an adjective, but there is no "gay" noun equivalent for "homosexuality." Perhaps most importantly, the word "homosexuality" can be put into action in diverse spheres of life. Most alternatives may work well in some contexts but not in others. While "isosexuality" could perhaps appeal to some zoologists, it would likely confuse nonprofessionals and annoy homosexuals. Conversely, "gay" may be appealing to many homosexuals and news organizations but not to scientists, who consider it too colloquial and too limited in scope.

For all these reasons, we have decided to stick to Kertbeny's "homosexuality" and to his broad interpretation of the term. We do, however, readily acknowledge that this decision will not resolve all terminological problems. One issue with our definition of homosexuality is that it seems to simply shift the initial question (What is homosexuality?) to another, possibly even more difficult question: What is sexuality? What makes certain activities and physical states sexual? When two men are kissing, are they having sex? When one man is watching another man masturbate, are they having sex? And what about mental states? What turns a desire into a sexual desire, for example? These questions will have to wait until chapter 2, where we discuss the concept of sexuality in the context of homosexual behavior in nonhuman animals.

Another issue with our definition of homosexuality is that applying the term can result in confusion. Suppose, for example, that you are a man and that you are sexually attracted to Kim. Kim is a biological man who self-identifies as a man, but you believe Kim is a biological woman, and you are attracted to Kim *because* you believe Kim is a biological woman. Suppose that your attraction would be over as soon as you found out that Kim is a biological man. Is your attraction a homosexual attraction? Philosophers make a distinction between *de dicto* and *de re* desires or attitudes to address such issues. Based on a *de dicto* interpretation, your attraction to Kim is not homosexual. You are not attracted to Kim because he is a biological man or because Kim self-identifies as a man. Based on a *de re* interpretation, the attraction *is* homosexual because you are nonetheless attracted to a biological man who self-identifies as a man.

These issues may not be hugely important for the natural sciences that study homosexuality. However, the distinction between *de re* and *de dicto* can be important for the social sciences and the humanities, not because they are all profoundly interested in the metaphysics of sexual orientation but because they focus more on intra- and intercultural variation in sexual behavior and desire. Hence, they try to answer questions such as whether people feel attracted to the biological sex of other people or to the associated gender (Sandfort 2005), or even to something totally different, such as bodily characteristics (other than sex), personality types, social status, or age (for a lengthy discussion of this issue, see Stein 1999, 61–67).

WHAT THIS BOOK IS *NOT* ABOUT

Even though this book covers a lot of ground, it has its inevitable limitations, two of which deserve some attention.

Our first limitation is that we focus primarily on male homosexuality, both in humans and in other animals. This book is not about female homosexuality. There are two reasons for this focus.

First, there is simply more information about male homosexuality because both scientists and philosophers have always given it more attention than female homosexuality. The philosopher Arthur Schopenhauer (2018, 577) once sneered that, on reading Socrates's orations, "one would almost come to believe that women simply don't exist." Part of the explanation for this bias is that most scientists and philosophers were men themselves, while women were often restricted from both professions. Another part has perhaps to do with the rather curious—but related—fact that male homosexuality has always kicked up more dust than its female version. According to Thomas Laqueur (1992, 53), Greek, Roman, and early Christian authors wrote much more about male homosexuality than about female homosexuality "because the immediate social and political consequences of sex between men [were] potentially so much greater."

Second, what we do know about female homosexuality seems to set it apart from its male counterpart. Recent research shows that there are numerous biological and other differences between human male and female homosexuality (for an overview, see J. Bailey et al. 2016). On a behavioral level, for example, human male homosexuals tend to be more sexually active than female homosexuals, at least in the West. The latter's sexuality has also been shown to be more fluid (again, at least in recent Western populations), with more women engaging in both homosexual and heterosexual activities.[7] On an etiological level, some biological factors involved in male homosexuality do not seem to play any role in female homosexuality. The fraternal birth order effect, for instance, which we discuss at length in the first and third chapters of this book, only seems to apply to male homosexuality. This effect holds that the number of older brothers is a good predictor of the sexual orientation of their later-born brothers but not their sisters. Interestingly, some scientists have even argued that it is a typically Western mistake to lump together male and female homosexuality (S. Murray 2000).

A second limitation of this book is that it highlights some but not all of the gay sciences. The reader will find that we take a lot of time discussing findings, concepts, arguments, and theories from comparative zoology, psychiatry, anthropology, evolutionary biology, social psychology, developmental biology, and even machine learning. Conversely, we pay less attention to developments in endocrinology, behavioral genetics, and neuroscience, even though these disciplines have perhaps been as popular and as productive in the past decades. It is indeed important to mention that our focus is based not on judgments about the value of these sciences but rather on our expertise, our idiosyncratic interests, and our intellectual background—the history and philosophy of science approach discussed earlier in this introduction. Indeed, some sciences simply fall outside the area of expertise that we have marked out for ourselves in the past fifteen years. Others are excluded because they already played a principal part in previous philosophical books on science and homosexuality, including Edward Stein's unsurpassed *The Mismeasure of Desire* (1999). Moreover, as philosophers of science, we have always been more interested in fostering dialogue among the sciences, particularly if they belong to different science groups, and this broad interest comes at the cost of comprehensiveness.

WHAT THIS BOOK IS ABOUT

This book is about the history and philosophy of some of the sciences studying human male homosexuality. It consists of four chapters and an epilogue, each of which can be read separately.

In chapter 1, we review some of the core concepts used by both philosophers and scientists to talk about the innateness and naturalness of homosexuality. We start (again) with the frank questions posed by the author of the *Problemata* in late antiquity, and we discuss how his questions and answers shaped part of the medieval debate on homosexuality's place in God's plan inscribed in nature. We then go on to sketch how these philosophical and theological debates were subtly transformed during the course of the nineteenth and twentieth centuries into scientific debates about the nature and causes of homosexuality. Finally, we scrutinize the arguments and the conceptual machinery used by so-called nativists, who claim that homosexuality is innate, and by their opponents, the infamous social constructivists. We expose the concepts used by both nativists and social constructivists as rather hopelessly confused and confusing.

In chapter 2, we examine whether there is such a thing as animal homosexuality. We begin with a reconstruction of one of the first scientific debates about the phenomenon in question: the debate about homosexual copulation in cockchafers or maybugs (*Melolontha vulgaris*) in nineteenth-century European entomology. The maybug story serves as a springboard for a philosophical discussion about a number of aspects of animal homosexuality—a discussion that reverberates over the other chapters of this book. These aspects include animal homosexuality's naturalness, its underlying determinants, its multidimensionality, its moral implications, and, last but not least, its relationship with human homosexuality.

Chapter 3 is a philosophical exploration of the question whether and how homosexuality has evolved. Again, this question touches on the issue of naturalness. Some argue that homosexuality is natural because it is a product of natural selection; others believe it to be unnatural because it is a nonreproductive form of sexuality seemingly inconsistent with the basic principles of the process of natural selection. We begin with a very brief history of how homosexuality came to be seen as an evolutionary paradox. We also discuss some solutions to this paradox. We then deconstruct the paradox by questioning some of its central assumptions, including claims about the homogeneity of human male homosexuality and about its impact on reproductive success. We conclude with a critical analysis of two recent evolutionary hypotheses about human homosexuality that manage to avoid these assumptions.

Chapter 4 is closely related to the previous chapters as it discusses psychiatry's dealings with homosexuality. While some psychiatrists appealed to its naturalness or innateness in their campaigns to normalize homosexuality, others appealed to its alleged unnaturalness to pathologize it. The chapter opens with an extensive psychiatric history of homosexuality, which we hang on two of its milestones: the German psychiatrist Richard von Krafft-Ebing's *Psychopathia sexualis* (1886) and the consecutive editions of the American Psychiatric Association's *Diagnostic and Statistical Manual of Mental Disorders* (*DSM*). We then use these historical explorations to inform our analysis and critique of the philosophical assumptions and implications of the claim that homosexuality is (or is not) a mental disorder. What do we actually mean when we consider a condition, such as homosexuality, a disorder? And what is actually wrong, either theoretically or morally, with saying that homosexuality is a disorder?

The epilogue builds on the four chapters to explore the claim that some kinds of research on homosexuality, particularly in the biological and biomedical sciences, should not be performed because such research stirs up the stigmatization of homosexuals. The main argument for this claim seems to be that there are strong ties between beliefs about the biology of homosexuality and attitudes toward homosexuals. However, recent research in social psychology adds nuance to that argument and complicates the relationships between attitudes and beliefs, particularly in the context of sexuality. This complexity can be illustrated with an anecdote about the American singer and anti-gay campaigner Anita Bryant. In an interview with *Playboy* in 1978, Bryant claimed that "even barnyard animals don't do what homosexuals do." When the interviewer kindly contradicted her with some relevant evidence, she retorted: "That still doesn't make it right" (Kelley 1978, 82).

Bryant was of course not the first individual to sport negative attitudes toward homosexuality and homosexuals, that is, homonegativity.[8] If anything, our book reveals how, throughout the ages, many philosophers and scientists have suffered from the same flawed perspective. In the 1950s, the German-American medical geneticist Franz Kallmann, one of the protagonists of chapter 1, considered homosexuality to be "an inexhaustible source of unhappiness, discontent, and a distorted sense of human values" (Kallmann 1952a, 296). Chapter 4 details how and why, well into the second half of the twentieth century, many psychiatrists supported the view that human homosexuality was a miserable mental disorder. In retrospect, the various conversion therapies on which they capitalized were not only ineffective but also positively harmful. Moreover, all the nineteenth-century entomologists lined up in chapter 2 clearly abhorred any kind of homosexuality. Echoing Pietro d'Abano's moral condemnation of human homosexuality at the beginning of this introduction, they described both human and animal variants as "morbid," "monstrous," and even "hideous." And even as recently as 1987, the *Entomologist's Record* published a paper on homosexual behavior in a butterfly species, in which the author complained that "national newspapers" were "all too often packed with the lurid details of declining moral standards and of horrific sexual offences committed by our fellow *Homo sapiens*" (Tennent 1987, 81).[9] The first evolutionary biologists to study homosexuality (in the 1960s), whom we discuss in chapter 3, however, did manage to avoid the persistent pathologizing and moralizing of their peers. Still, some commentators accuse contemporary evolutionists (and other scientists) of using "pathological sounding language," and thus contributing to "an implicit [negative] value judgment as to homosexuality's worth" (Roughgarden 2017, 503, 505).

All in all, however, it seems fair to say that there is very little homonegativity in today's gay sciences, compared both to their own history and to current public opinion, part of which still struggles with homosexuality. Some scientists openly dedicate themselves to promoting gay liberation, even though many realize that their own work may not always be useful for that purpose. It is nevertheless true that talking about homosexuality will increase its familiarity, and increasing its familiarity will likely reduce homonegativity. We can only express our hope that this book also contributes to that good cause. As Edward Stein (1999, 5) said, "The love that dare not speak its name is now unwilling to keep quiet." And so it should be.

[CHAPTER 1]

Not by Genes and Hormones Alone

On Homosexuality and Innateness

Quoniam nullum naturale turpe [est].

WALTER BURLEY

Naturam expellas furca, tamen usque recurret.

HORACE

Are homosexuals born that way? Lady Gaga seems to think they are, and so do likely many of our readers. Gallup has polled the American people on this question for over forty years. In the mid-1970s, only 13 percent believed homosexuality to be innate, while a clear majority associated it with "upbringing and environment," as the poll phrased it (Saad 2018). These views are often described as *nativism* and *environmentalism*, respectively. Since 2015, however, a majority of Americans' thinking about homosexuality has become nativist, with the proportion of environmentalists falling to a historical low of 30 percent. Recent large-scale public opinion polls on this question have not been conducted in other Western countries, but it seems that many had already experienced a similar shift in opinion long before the United States (De Boer 1978).

Why has popular opinion changed so drastically over the past four decades? If we ask our rather well-educated friends, some speculate that the shift is due to new scientific evidence that has surfaced only recently. Others connect it to an increasing scientific literacy among the public, while still others bet on a combination of such factors. All these explanations assume that science now tells us with certainty that homosexuality is innate. Moreover, there is an even deeper assumption that the issue itself *can* be solved in principle by science.

However, a number of contemporary philosophers question these assumptions, and we share their skepticism. In our view, the question whether homosexuality is innate cannot be meaningfully answered by science. Surely, future scientific research will give us an ever-more accurate view of the causes and mechanisms of human male homosexuality, but we are doubtful about the possibility of mapping these findings onto the innate-acquired spectrum, mainly because we believe that thinking about this issue along such lines is too confused to be productive. As always, other philosophers disagree, but even if they are right, the onus is on them to show that there is solid scientific evidence to make a case for (or against) nativism.

We start this chapter with three sections that outline a brief history of the innateness debate. The fourth section analyzes what contemporary philosophers think we mean when we ask whether some trait is innate and focuses on two philosophically and scientifically informed answers to this question. We then show that innateness claims about homosexuality cause so much confusion that it is probably better to avoid endorsing or denying them. Finally, we argue that the same conclusion holds for anti-nativist or so-called social constructivist claims about homosexuality.

MOIST BUTTOCKS IN THE *PROBLEMATA*

Since antiquity, philosophers have pondered the naturalness of male love. One of the boldest reflections on this subject can be found in the *Problemata*, a compilation of ancient Greek natural philosophical texts, with which this book began.[1] In total, the work contains almost one thousand questions, all of which relate to natural phenomena and range from drunkenness, bruises, "shrubs and vegetables" (book 20), barley flour, and "malodorous things" (book 13) to the color of the skin. The text then outlines a solution for each problem, usually by referring to natural causes.

Book 4 of the *Problemata* is entirely devoted to questions about sexual intercourse and sexual pleasure. From a contemporary perspective, some questions and most of the answers are positively bizarre, mainly because they are based on humoral pathology, the doctrine of bodily fluids. Humoral pathology holds that our body has four different kinds of humors: black bile, yellow bile, blood, and phlegm. Black bile was thought to be cold and dry; yellow bile, hot and dry; blood, hot and wet; and phlegm, cold and wet. Humoral imbalances supposedly led to all kinds of illnesses and oddities. In the *Problemata*, for example, the author wonders why both birds and hairy humans are lustful. The answer is succinct: "Because they contain much moisture" (book 4, question 31). In our general introduction, we mentioned that the book also contains frank questions about the nature and causes of homosexuality and quoted the question: "Why do some men enjoy submitting to sex, and some at the same time enjoy being active, whereas others do not?" (book 4, question 26). The question itself indicates that the ancient Greeks already distinguished different forms of male homosexuality based on the preferred sex act: some men prefer to be penetrated, whereas others alternate between penetrating and being penetrated. (This distinction will continue to surface in the next chapters, as in, for example, the context of animal models of human homosexuality in chapter 2.) The author of the *Problemata* focuses primarily on the first, "passive" form of homosexuality and suggests two possible explanations.

A first possibility is that "passive" homosexuals have an anatomical abnormality, as a result of which their semen chiefly accumulates not in the testes and the penis but in their buttocks. This accumulation causes an urge for friction, which then shows itself as a desire to be penetrated. The underlying anatomical abnormality can be either innate or acquired at a later age, such as with eunuchs. The first case concerns men who, according to the author, are "effeminate by nature" (book 4, question 26). In less feminine men, semen collects in various places, which makes these men able to perform multiple sexual roles. A second explanation for "passive" homosexuality is that it results from habit. According to the author, those men who regularly had passive sexual relations with older men during puberty (a form of homosexuality that we described as *transgenerational* or *ritualized homosexuality* in our general introduction, and which we will discuss in more detail in chapter 3) would also prefer this form of sexual intercourse at a later age. In some cases, the desire to be penetrated becomes so dominant that "the habit becomes more like a nature" (book 4, question 26)—a second nature, as it were.

Medieval scholars only discovered the *Problemata* in the late thirteenth century, after Bartholomew of Messina translated it into Latin around 1250. In the meantime, however, sexual morality had changed dramatically (Cadden 2001, 2003; Wallis 2008). Even by today's standards, the Romans and especially the ancient Greeks were quite tolerant of all sorts of sexual activities. It is certainly no coincidence that "Greek love" is a common synonym for "homosexuality." The advent of Christianity altered all this. Already in the fourth century, Christian emperors imposed the death penalty for homosexuality (C. Williams 2010), but from the eleventh century onward, ecclesiastical authorities started a real crusade against homosexual acts and relationships. Thus the Aristotelian question and its answers set off a time bomb in an intellectual world that had come to consider homosexuality both a crime and a mortal sin, if not a serious form of blasphemy. Understandably, many late medieval Christian commentators of the *Problemata* gave the topic a wide berth, while those who had the guts to deal with it felt compelled to express their aversion in clear terms.

One of the latter was the fourteenth-century philosopher Pietro d'Abano. Unlike the *Problemata*'s author, he utterly condemned all homosexual acts as perverse, filthy, wicked, and even monstrous (cited in Cadden 2001, 75n32). At the same time, d'Abano insisted that natural philosophers should study the subject, rather than ignore it, as many of his contemporaries did. After all, following the *Problemata*, he believed that certain forms of "passive" homosexuality were part of, or arose from, the nature of the individual (Coucke 2009). Hence, natural philosophers could and must look into the issue, no matter how horrible they found it. (When a later commentator, the theologian Walter Burley, emphasized that "nothing natural is shameful"—one of the epigraphs to this chapter—he was referring to the *study* of homosexuality, not to the phenomenon itself.) According to d'Abano, some homosexuals are born that way, either as an inevitable consequence of an innate anatomical abnormality, as the Aristotelian author had suggested, or because of the position of the planets at the time of their conception. According to d'Abano, this individual nature provides an explanation for their desires and behaviors. That nature is nevertheless monstrous, because, d'Abano claimed, it goes against the nature of the human species in being nonprocreative.

D'Abano believed that some forms of "passive" homosexuality were innate, and therefore unchangeable. He agreed with the medieval physician Avicenna on this point, who thought it was ridiculous to even want to cure all forms of homosexuality ("Et stulti homines sunt qui volunt eos curare"; quoted in Cadden 2001, 76). However, d'Abano disagreed with Avicenna on the cause of homosexuality, which Avicenna believed to be "meditative" rather than innate ("Nam initium egritudinis eorum meditativum est non naturale"; quoted in Cadden 2001, 76). In Avicenna's view, one should look for the cause of homosexuality not in the body but in the mind, and more specifically in the imagination. Avicenna and d'Abano also differed on how to alter those acquired forms of homosexuality that they did consider changeable. According to Avicenna, one could only try to break some homosexual desires, for example, by starvation, imprisonment, or some serious spanking. D'Abano's therapy, however, focused primarily on somatic interventions. To exhaust sexual desires, he recommended medication and a diet.

Interestingly, the Christian philosopher d'Abano had no difficulty recognizing that his analysis seemed to imply that (some) homosexuals cannot be held responsible for their actions (Cadden 2001, 77). After all, one cannot change nature. One may try to expel it, but like Horace (quoted in the second epigraph to this chapter), d'Abano believed it would surely return sooner or later.

INNATE? NATURALLY!

In ancient, medieval, and early modern philosophies, countless hypotheses circulated about the causes of the various forms of homosexuality. It is not our intention to review them here (for an overview, see Cadden 2003 and Borris and Rousseau 2008). However, we do want to draw the reader's attention to their conspicuous continuity with more recent scientific explanatory hypotheses and to the deep polysemy of some of the terms they use, including "natural," "constitutional," and "innate."

The first champions of gay rights in the nineteenth and twentieth centuries made extensive use of such terms. These activists included Károly Kertbeny and Karl Heinrich Ulrichs, whom we mentioned in our general introduction, and the physician Magnus Hirschfeld, who will also reappear in chapter 4. In fin de siècle Berlin, Hirschfeld argued for a scientific approach to human sexuality in the hope that the new science of sexology would promote the emancipation of male and female homosexuals. His motto was *Per scientiam ad justitiam* (From science to justice; see fig. 1.1, proposition 14).

Hirschfeld followed Ulrichs in claiming that there were scientific grounds to believe that homosexuality was natural and innate, and that it was therefore absurd to prosecute homosexuals, as was still done in Germany and many other countries at the time. Legally, homosexuality was often defined as a kind of sexuality that went against nature—the German law in question spoke of "widernatürliche Unzucht" or "indecency against nature." Hirschfeld, however, believed that, at the very least, homosexuality was not against the *individual* nature of homosexuals: "[Many] findings and experiences... have led homosexuals to conclude that the legal provisions on *widernatürliche Unzucht* do not bear on them at all. After all, their interaction does not go against nature, or so they argue, at least not against their own nature, which attracts them to people of the same sex. However, every individual must be judged 'ex natura sua' [by their own nature], and not on a nature that is alien to him or her" (Hirschfeld 1914, 313).

Hirschfeld even went one step further. Unlike many ancient and medieval commentators, he did not think homosexuality was at odds with the nature of the species either. Here he followed Ulrichs's logic again, basing his claim on the then scarce findings about the occurrence of homosexuality in the animal world. (For more on this, see the next chapter.)

In his 1914 magnum opus, *Die Homosexualität des Mannes und des Weibes* (The homosexuality of men and women), Hirschfeld devotes two chapters to the idea that homosexuality is natural and innate, two terms he appears to consider synonyms (Hirschfeld 1914, chaps. 17 and 18). His list of arguments

**Leitgedanken und Sinnsprüche
im Institut für Sexualwissenschaft.**

Um die Grundgedanken des Instituts zu veranschaulichen, verfaßte Dr. M. Hirschfeld für den Ernst Haeckel-Saal folgende L e i t s ä t z e :

1. Die Nächstenliebe erfordert, die Liebe des Nächsten zu achten.
2. Statt: „Wer ist schuld?" fragt „Was ist schuld?"!
3. Jeder Mensch ist das Ergebnis seiner Anlage und Lage.
4. Wie jede Anziehung in der Natur beruht auch die der Liebe auf Gesetzen.
5. Kein Gesetz ohne Ausnahme, keine Ausnahme ohne Gesetz, bedingt ist alles.
6. Liebe ist Umsetzung ruhender in lebendige Kraft.
7. Wie Leben aus der Liebe sprießt, sproßt aus dem Leben Liebe.
8. Wer beiden Geschlechtern entstammt, enthält beide Geschlechter vereint.
9. Die Begriffe übernatürlich, unnatürlich und widernatürlich sind Zeichen mangelnder Naturerkenntnis.
10. Sitte wie Sittlichkeit wechseln nach Ort und Zeit.
11. Der nackte Mensch ist nicht ausgezogen sondern nicht angezogen.
12. Freiheit verpflichtet.
13. Vorurteile sind Nachurteile.
14. Per scientiam ad justitiam (durch Wissenschaft zur Gerechtigkeit).
15. Die Wissenschaft ist nicht wegen ihrer selbst, sondern um der Menschen willen da.
16. Die wahre Reinheit ist die reine Wahrheit.
17. Über alles die Wahrheit.

Figure 1.1: Some basic principles held by the Berlin Institut für Sexualwissenschaft, directed by Magnus Hirschfeld (1924). Source: Institut für Sexualwissenschaft, Dr. Magnus Hirschfeld-Stiftung 1924, 34.

was based primarily on a large-scale survey of one thousand homosexuals about all aspects of their sexual life, a "psychobiological questionnaire" he conducted himself. These are Hirschfeld's eight main arguments for the innateness of homosexuality:

1. Homosexuality is innate because it is still expressed in environments that keep it under wraps, consider it as horrible and even monstrous, or try to "contain" or "eradicate" it. From this persistence, one can infer that homosexuality is indeed a deep-rooted property and not some fashionable phenomenon.

2. Homosexuality is innate because it already manifests itself in young children in the form of "girlish" traits in boys and "boyish" traits in girls. Hirschfeld speculated that some homosexuals constitute in fact a kind of "third sex," having both female and male sex characteristics.

3. Homosexuality is innate because almost all homosexuals later recall that their very first sexual fantasies and desires, and their very first wet dreams, related to an individual of the same sex. Homosexual desires are present from the early awakening of sexuality.

4. Homosexuality is innate because it permeates the entire being of the individual. Half a century later, the French philosopher Michel Foucault would formulate the same thought: "Nothing that went into his [the homosexual's] composition was unaffected by his sexuality. It was everywhere present in him: at the root of all his actions because it was their insidious and indefinitely active principle; written immodestly on his face and body" (Foucault 1978, 43).

5. Homosexuality is innate because it is a stable and unchanging characteristic. Here Hirschfeld's underlying reasoning was that if homosexuality were not innate but rather acquired through external circumstances, then other circumstances should suffice, at least in principle, to make it disappear again. In actual practice, however, Hirschfeld noted it is impossible to "convert" homosexuals to heterosexuality, and vice versa.

6. Homosexuality is innate because heterosexuality and other sexual orientations are also innate. There does not seem to be a substantial difference between the outward appearance of homosexual and heterosexual love ("the search and yearn; joy and sorrow," as Hirschfeld [1914, 318] put it succinctly), so why would there be a difference when it comes to their underlying causes?

7. Homosexuality is innate because it is universal. According to Hirschfeld (1914, 322), the phenomenon occurs "in all ages, in all regions, in all peoples, in all professions and in all stages of civilization . . . , as well as in the animal world and the vegetable kingdom." In his view, this ubiquity implied that homosexuality is part of human nature. On this point, Hirschfeld quoted a

long passage from Arthur Schopenhauer's *Die Welt as Wille und Vorstellung* (The world as will and representation). It is well known that Schopenhauer was a misogynist, but his homonegativity is perhaps less well known. "In itself," Schopenhauer (2018, 577) believed, "pederasty [or homosexuality] presents itself not merely as an unnatural monstrosity but also as disgusting and repulsive to the highest degree, an act to which only a completely perverse, distorted, and degenerate form of human nature could ever have descended, and that could recur in completely isolated cases at most." At the same time, Schopenhauer could not imagine a single era or culture in which homosexuality had not been reported, despite the continued repression and harsh punishment of homosexuals. His conclusion: "The complete universality and persistent ineradicability of the practice prove that it stems in some way from human nature itself" (578). Schopenhauer thus agreed with Hirschfeld that the universality of homosexuality is a proof of its innateness.

8. Homosexuality is innate because it seems to run in families. According to Hirschfeld, his figures showed that nearly a quarter of his homosexual subjects knew of one or more homosexual close relatives. In his days, genetic research on homosexuality was still in its infancy, as we will see in the next section, so Hirschfeld was limited to a few rather amusing anecdotes to prove his point. For example, he quotes from the testimony of "a skilled writer" and homosexual who told him that his father's mother was "rather masculine in her way of doing things, and developed a beard late in her life" (Hirschfeld 1914, 320). Another testimony reads almost like a joke:

> A gentleman from a small city one day visited an older homosexual in Munich. He came on the recommendation of a common friend from the province. The old man liked the visitor, and it didn't take long before the old man made a pass. The visitor, however, resisted and said, "I would have imagined you much younger based on my friend's description." "Oh," replied the older man, "then it might have been about my son, who is also like me [*der ist auch so*], but unfortunately he is traveling." (322)

Hirschfeld's list of arguments is mostly very traditional and clearly in line with the many arguments that had already been offered in ancient and medieval Europe. It also, however, features new arguments that remained popular throughout the twentieth century and even into the twenty-first century. This is particularly true for the argument that considers homosexuality as innate because it runs in families. In the next section, we will delve into the early twentieth-century genetic research on sexual orientation and explore how this research was used to argue for the innateness of homosexuality.

SCHIZOPHRENIC FAMILIES AND HOMOSEXUAL TWINS

The rise and fall of various scientific disciplines has produced different answers to the question whether homosexuality is natural or innate. For Aristotelians, humoral pathology provided many explanations, whereas Hirschfeld never once mentioned bodily fluids in his discussion of the origins of homosexuality. In twentieth-century Europe, most people seemed to assume that the answer should come from the sciences, such as psychology or genetics. If homosexuality is in our genes, they reasoned, then it is clearly innate.

Hirschfeld had already noted that homosexuality runs in families, but he did not have any reliable data to substantiate his hunch. It wasn't until 1952 that the first more or less systematic study of the genetic aspects of homosexuality was published. The paper was written by Franz Kallmann, a psychiatrist and geneticist who started his scientific career as one of the many employees of the world-renowned Deutsche Forschungsanstalt für Psychiatrie (German Institute for Psychiatric Research) in Munich. Founded in 1917 by the famous psychiatrist Emil Kraepelin (Weber 2000), the institute's goal was to uncover the many physical causes, including genes, that contributed to several conditions that were considered mental disorders.

The interest in the genetic causes of mental disorders was in a sense a continuation of the biological-psychiatric research program that had already dominated the second half of the nineteenth century (see chapter 4). During the interbellum period, this interest was further fueled by German nationalists, including the National Socialists, and their concern for the future of the German people, who had been greatly weakened by the havoc and human losses of World War I and the Spanish flu pandemic (Weindling 1989). The state called in psychiatrists to devise measures promoting mental and physical health. One of those measures was to identify disorders, or conditions that were assumed to be disorders, that were believed to have a strong genetic anchoring, such as schizophrenia, manic depression (now known as bipolar disorder), alcoholism, homosexuality, and some forms of criminality. German psychiatric geneticists were tasked with finding scientific evidence for these presumptions.

At a later stage, their findings would play a crucial role in many eugenic measures, including a large-scale sterilization campaign for "hereditary patients" and their relatives. This campaign initially began in Germany with voluntary sterilization, but in 1933, a year after the National Socialists came to power, Hitler passed a law that made sterilization compulsory for the affected population. The purpose of this law and its later variants was clear:

to eliminate "hereditary disorders" with the help of medical interventions, so that later generations could be free from disease and crime. At least four hundred thousand individuals, including many homosexuals, were sterilized on the basis of this law (Proctor 1988, 109). And it only got worse. In the chaotic build-up to World War II, the German government escalated from forced sterilization to the murder of children and adults with all kinds of hereditary disorders under the (secret) Aktion T4 campaign. Recent figures estimate that at least three hundred thousand people were killed in this campaign: children, adults, and elderly people with a mental or physical disability; homosexuals; psychiatric patients; and later sick forced laborers and civilians traumatized by the war. Over the past decade, German historians have mapped out this horror, recovering documents that were long considered lost (Rotzoll et al. 2010). Many homosexuals also died in concentration camps, and those who managed to avoid death often only did so by agreeing to castration (Plant 2011).[2] Some gay concentration camp survivors were transferred directly to prisons to serve their remaining time.

Kallmann was affiliated with the Deutsche Forschungsanstalt between 1929 and 1935. During that period, he became a sort of academic star with his ambitious research project on the genetics of schizophrenia—a study that, eventually, would also lead him to look into the genetics of homosexuality. In his first schizophrenia study, he identified no less than 1087 individuals with schizophrenic symptoms and constructed a family tree for each of them with details about the psychological health of all family members: partners, parents, brothers and sisters, children, and cousins (Kallmann 1938). The results were striking: one in six children of patients with a "non-schizophrenic" partner also became schizophrenic, compared to less than one in a hundred children in the rest of the population. Even cousins of patients were still four times more likely to develop the condition than individuals from non-schizophrenic families (145). Nearly three-quarters of the children of parents who were both schizophrenic became schizophrenic themselves (161). Kallmann's conclusion was that genetic components must play an important role in the development of schizophrenia, although he insisted that they do not tell the full story, since some family members of schizophrenic patients were non-schizophrenic.

In later studies, Kallmann shifted his focus from family studies to twin studies. Family studies suggested that genetic material played a role in the development of schizophrenia, but according to Kallmann they did not provide entirely conclusive evidence. It was, after all, quite possible that children of schizophrenic parents developed schizophrenia themselves not because of the genetic material they inherited but, for instance, because of their parents' own problematic choices or because of a specific and learned parenting style. Partly influenced by psychoanalysis, the hypothesis that parental attitudes are responsible for the mental disorders of their offspring was already very popular in American psychiatry before World War II, and it would even increase in popularity afterward (Shorter 1997). If Kallmann wanted to rule out this possibility, he needed more than just family studies: he also needed twin studies.

The focus on twins was further spurred by Kallmann's emigration to New York in 1936—a move necessitated by his Jewish heritage. In New York, he immediately noticed that his American colleagues were less than enthusiastic about his psychiatric-genetic research project. To convince them of the value and findings of his earlier research, he went looking for twins. Twins, he believed, were "living evidence ... for the impact of gene-specific processes on human behavior" (Kallmann 1953, 107). Parents and children share on average only half of their genetic material, while identical or monozygotic twins share virtually all their genes, except for some mutations that occur after fertilization. Monozygotic twins arise from the fertilization of a sperm cell and a single egg, while nonidentical (or dizygotic) twins result from the fertilization of two sperm cells and two ova. On average, nonidentical twins share the same number of genes as non-twin siblings. If schizophrenia is to a large extent a genetic disorder, Kallmann reasoned, then it is to be expected that both members of identical twins will experience the condition if one of them does; this will less frequently be the case with dizygotic twins (of the same sex), even though dizygotic and monozygotic twins largely share the same environment. The technical term for the degree to which both twins express a given trait, such as schizophrenia or homosexuality, is "concordance."

In one of his most famous studies, Kallmann presented data about the fates of 174 patients with schizophrenia who had an identical twin and data about their close relatives. The concordance figures were startling: 85 percent of identical twins both became schizophrenic,[3] while regular brothers and sisters, as well as dizygotic twins, had a "mere" 15 percent chance of developing schizophrenia if a sibling was schizophrenic. Kallmann (1946, 315) added that "over 85 percent of our groups of siblings and dizygotic twins did *not* develop schizophrenia, although about 10 percent of them had a schizophrenic parent, all of them had a schizophrenic brother or sister, and a large proportion shared the same environment with these schizophrenics before and after birth" (italics in original). In his study, Kallmann also explained that genetic material might be primarily responsible only for the susceptibility to schizophrenia. In most cases, the actual process of developing the condition is then triggered by all kinds of adverse life events, he believed, and many of the specific symptoms of and future prospects for patients are determined by all sorts of other environmental factors.

Kallmann was not only a pioneer in genetic research on schizophrenia. He was also the first scientist to conduct large-scale empirical research on the genetic aspects of homosexuality. As mentioned before, homosexuality and schizophrenia were among a series of disorders for which the scientific community of the time suspected a strong genetic cause. In a study that appeared a year before James Watson and Francis Crick published their groundbreaking paper on the molecular structure of DNA, Kallmann described such conditions as "genetically controlled imperfections" (Kallmann 1952a, 283). The unsuspecting reader might frown at the classification of homosexuality as a disorder, but as we will explain in chapter 4, many European and American psychiatrists vigorously defended the disorder view of homosexuality until well into the 1970s. Moreover, in many Western countries, homosexuality was considered not only a disorder but also a crime. Hence, almost all the test subjects in Kallmann's research came from psychiatric institutions, detention centers, and prisons. Kallmann himself traced a few more test subjects, all of whom lived on the fringes of society and had little or no contact with their family. To Kallmann's surprise, they were quite rebellious and unhelpful: "We are ready to concede at this point that an investigation of the sexual habits and self-protective devices of an ostracized class of people and their family relations is not a promising field of exploration for research workers who are in any way concerned about their conventional peace of mind" (Kallmann 1952a, 283).

In the end, Kallmann managed to collect data on forty identical male twins, forty-five dizygotic male twins, and their close relatives. The resulting study contains a number of double portraits of identical homosexual twins, which were made "unrecognizable" by a black bar that covered their eyes (see figures 1.2–1.4).

The results of Kallmann's twin study on homosexuality were even more remarkable than the results of his earlier studies on schizophrenia. Perhaps the most surprising finding was that *all* sets of identical twins were both homosexual—a spectacular 100 percent concordance. (In a later work, Kallmann [1953, 116] reported a single exception: the case of thirty-year-old identical twins, only one of whom was homosexual, schizophrenic, and alcoholic.) Dizygotic twins and ordinary siblings had hardly a greater chance to be (or to become) homosexual than random individuals from families without any (open) homosexuals. Despite these spectacular data, Kallmann remained very cautious in his conclusion. In his eyes, his data only showed that genetics was part of the explanation for the tendency toward (some forms of male) homosexuality. By themselves, genes are incapable of driving a man into the arms of another man, or so Kallmann claimed, but they do direct the sexual development of an individual in a certain direction,

Figure 1.2: Kallmann's "J. twins at age thirty-nine... who traveled as stewards all over the world, were sturdy and masculine in both their appearance and sex activities" (1952). Source: Kallmann 1952a, 292–93.

so that it becomes more likely—possibly much more likely—that this individual ultimately will become homosexual. Kallmann speculated that particular genes code for an underlying endocrine mechanism that plays a role in the process of psychosexual development, but he emphasized at the same time that the causality of the condition may turn out to be very complex and take place at different levels (Kallmann 1952a, 294).

Kallmann concluded that the evidence for a genetic basis of homosexuality was clearly suggestive but still not entirely decisive. He believed that conclusive evidence could only come from cell-biological (or molecular-genetic) research: "In the absence of such ... data, it is fair to admit that the question of the possible significance of genetic mechanisms in the development of overt homosexuality may still be regarded as entirely unsettled" (Kallmann 1952a, 286).

Some decades later, Kallmann's caution proved to be justified. First, contemporary concordance rates for male homosexuality have turned out to be dramatically lower than his rates, even though the monozygotic twin concordance rate still exceeds the dizygotic rate. In a recent paper, Gavrilets, Friberg, and Rice (2018, 27) summarize contemporary concordance studies and conclude that "estimates of proband concordance among twins (i.e., the probability that a twin is homosexual given that the other twin is homosexual) are low in both sexes: around 20 percent for monozygotic twins,

Figure 1.3: Kallmann's "O. twins" at age thirty-five. He described them as "heavy-drinking night club entertainers, who specialized as female impersonators and belonged to the entirely passive type of homosexual" (1952). Source: Kallmann 1952a, 293.

Figure 1.4: Kallmann's "K. twins" at age twenty-two. According to Kallmann, they "earned their living as models and entertainers until their careers were ended by excessive alcoholism" (1952).
Source: Kallmann 1952a, 293–94.

with smaller percentages for dizygotic twins." Second, Kallmann would probably be disappointed, though not entirely surprised, to hear that it still remains particularly difficult to identify the genes that might be involved in being gay, despite the zeal of many scientists who have devoted much of their career to this issue.

In the 1990s and the first decade of this century, the hunt for "gay genes" was covered in a series of high-profile publications (Hamer et al. 1993). Unfortunately, many of these so-called linkage studies, which focused on patterns of DNA sharing within families, contained conflicting results, and they were often difficult to replicate (Mustanski et al. 2005). Due to newly developed techniques, the last decade has seen a number of genome-wide association studies (GWAS) that scan the full genome of thousands of individuals to find genetic variants associated with homosexuality (Sanders et al. 2015, 2017; Ramagopalan et al. 2010). In 2019, a huge GWAS with almost thirty thousand homosexual individuals and more than four hundred thousand controls revealed several such genetic variants, some of which seem specific for male homosexuality (Ganna et al. 2019). Yet, each of these variants only has a minor effect on homosexuality, and taken together, these genetic factors explain less than 30 percent of the variation in sexual orientation, revealing the influence of genes on homosexuality to be moderate at best. As one recent review aptly puts it: "'Is sexual orientation genetic?' Th[e] answer is: 'Probably somewhat genetic, but not mostly so'" (J. Bailey et al. 2016, 76).

This conclusion raises a series of questions. If genes are not a major factor in explaining homosexuality, then what is? The environment? And if it is the environment, then does that mean that homosexuality is less innate than usually assumed?

CAN SCIENCE MAKE SENSE OF INNATENESS?

Though Kallmann himself never used the term "innate"—he preferred related terms, such as "natural," "organic," "genetic," "inheritable," and "constitutional"—some commentators have hailed his twin studies as evidence for Hirschfeld's hunch that someday science would prove homosexuality to be innate. However, Kallmann and Hirschfeld were nativist mavericks in a twentieth century dominated by environmentalist psychoanalytic thinking, and Freud and later psychoanalysts conceptualized homosexuality as a trait acquired in early childhood, rather than as a "genetically controlled imperfection" (Kallmann 1952a, 283). One of Freud's own ideas was that some kinds of homosexuality were caused by excessive identification with the mother because of either castration anxiety or the (relative) absence of a fatherly authority. (For more on this, see chapter 4.) And even Freud's main rival, the German psychiatrist Emil Kraepelin, generally considered the founding father of biological psychiatry, subscribed to the view that homosexuality could be induced by seduction (Mildenberger 2007). In concentration camps, the Nazis tried to isolate homosexual prisoners from other inmates because they were afraid that homosexuality was contagious (Lautmann 1980).

Today, Kallmann's position is much more fashionable, as the search for gay genes and hormones has led to high-profile papers that are eagerly quoted and discussed in the popular press. Remarkably, however, the process of popularization often transforms the very same results into wildly divergent conclusions. An excellent example is the recent genome-wide association study mentioned above (Ganna et al. 2019). Featured by news media all around the globe, the study compared the genomes of a homosexual group and a heterosexual control group and revealed five important genetic variants associated with homosexuality. According to some publications, like *Forbes*, the study showed that "being gay is 'natural'" (Ennis 2019). Others, like the Catholic News Agency, read the article differently and believed that it indicated that any attempt to reduce homosexuality to some so-called gay genes was doomed to fail (Condon 2019). The Ugandan newspaper the *Independent* even wrote that "Uganda's Minister for Ethics and Integrity Fr. Simon Lokodo has welcomed results of a recent massive study that shows that the so-called 'gay gene' is a total myth." The op-ed's title put it even more sharply: "Homosexuality not genetic, says study" (Akankwatsa 2019).

Ideological spinning is only part of the explanation for this notable divergence of interpretations. Intellectual confusion over the concept of innateness also plays an important explanatory role. Both laypeople and scientists tend to consider this concept scientifically useful in that it refers to a property that is relevant in scientific work. The only problem is that different individuals, whether scientists or laypeople, seem to have different views about just what that property is. Obviously, this is a big issue, because we can only say that science has shown homosexuality to be innate if we can at least agree on what property (or set of properties) homosexuality must have in order to qualify as innate.

In cases of conceptual confusion, one can always call on philosophy for a resolution. Philosophers have indeed tried to come up with accounts of innateness that save the concept's scientific value. As we will explain later in this chapter, we are skeptical about the success of these analyses and proposals. (This is a philosophy book, after all.) However, we still think it is worthwhile to discuss two of these accounts and then consider for both of them whether scientific evidence allows them to conclude that homosexuality is indeed innate.

Before we begin, however, the reader should know two things about the innateness debate in philosophy. First, all innateness accounts assume that a trait cannot be innate if it is learned. Philosopher Matteo Mameli (2008) considers this to be the minimal condition that any account of innateness must meet. It follows that objections to innateness accounts often revolve around examples of traits that have the property assumed to be the property that "innate" refers to but are nevertheless (partly) learned. Second, all innateness accounts seem to cluster around one (or more) of the following three features: species typicality, teleology, and fixity (Griffiths 2002; Griffiths, Machery, and Linquist 2009). A *species typical* human trait is a trait that typifies our species in the sense that it can be found among almost all members of our species. For this reason, species typicality is largely irrelevant in the context of homosexuality. Admittedly, some scientists and other intellectuals, including Freud, have entertained the idea that all human beings are born bisexual, but such ideas only play a minor role in today's scientific publications and media reports about the innateness of homosexuality. Moreover, if people really thought that homosexuality could only be innate if all or most members of our species were homosexual, then they would not believe that homosexuals are "born this way." *Teleology* has more importance for innateness discussions about homosexuality. A trait is teleological if it has a function for which it has been naturally (or sexually) selected, and some scientists do indeed believe that particular aspects of (some forms of) homosexuality are evolved adaptations. We discuss some of these adaptationist explanations in chapter 3.

In this particular context, however, *fixity* is the most relevant cluster of features that are supposed to be captured by the concept of innateness. According to Griffiths, Machery, and Linquist (2009, 609), a trait is fixed if it "is hard to change; [if] its development is insensitive to environmental inputs in development; [and if] its development appears goal-directed, or resistant to perturbation." Many people think of homosexuality as something that science has shown to be very hard to change, and they call it "innate" for precisely that reason. The arguments of both Hirschfeld and d'Abano nicely illustrate this line of thinking, and it can also be inferred from the fact that the many websites devoted to "curing homosexuality" appear to explicitly deny that it is innate.[4] After all, curing is a way of altering characteristics. Because fixity accounts of innateness are most relevant in the case of homosexuality, we will focus on two of these below: heritability and canalization.

Innateness as Heritability

Mendelian conditions are conditions caused by a single gene. Cystic fibrosis and Huntington's disease are two examples. If you have the gene for either of these conditions, you will eventually develop the disease, provided, of course, that you live long enough. Sadly, both Huntington's and cystic fibrosis are not currently curable, even though there are pharmacological and other therapies that can alleviate some of their symptoms. Unlike Mendelian disorders, most human conditions are influenced by many genes. Athletic ability, for instance, is associated with a large number of genes. Having particular genetic variants increases your chances of becoming a good athlete, but it does not suffice: diet and exercise are still important if you really want to succeed as a professional athlete. Genetically speaking, homosexuality is more akin to athletic ability than it is to cystic fibrosis. Recent genome-wide association studies (Ganna et al. 2019; Sanders et al. 2017) have identified a number of genes that increase the odds of developing a homosexual orientation.[5] Such findings, however, do not preclude the possibility of someone who is heterosexual despite having all the relevant "homosexual" genes. The question is, then, how much of our sexual orientation is determined by our genes (Roughgarden 2017). When answering this question, scientists often say that homosexuality is "mostly" or "moderately" in our genes, which suggests that they are talking about homosexuality's heritability.

Interestingly, heritability is one measure that has often been used to define innateness (Jacobs 1981), as it can tell us something about the fixity of a trait. It does indeed make sense to claim that a highly heritable trait is only minimally mutable by changing the environment (within the examined range of environments). But first things first: what exactly does heritability measure?

Heritability is a population measure: it expresses to what extent a population's interindividual variation of a trait, such as homosexuality, is determined by genetic variation, rather than by environmental variation. For a phenotypic trait with a heritability of 80 percent (or 0.8), only 20 percent of its variation can be changed by tweaking the environment. A heritability of 100 percent (or 1) indicates that all phenotypic variation is determined by genotypic variation; zero heritability (or 0) means that none of the variation is determined by genotypic variation and that environmental variations underly all the variation in the trait.

Of course, it often matters which type of environmental factors underly interindividual differences. For that reason, heritability studies distinguish between *shared environment* and *nonshared environment*. Elements of the shared environment are those elements that people growing up together will share because they grow up together. Nonshared environment is the sum of all environmental influences that are unique to an individual, such as accidents and contact with peers. Heritability studies attempt to determine a percentage for the contribution of each of these three factors: genes, shared environment, and nonshared environment.

To determine a trait's heritability, behavioral geneticists mainly use twin studies, much like Kallmann. They compare the concordance rate of homosexuality in monozygotic twins with the concordance rate in dizygotic twins. If genes are responsible for the phenotypic variation, the concordance rate should be higher in monozygotic twins, given that they share the same genes. If the shared environment between twins is important, concordance rates should be high for both monozygotic and dizygotic twins who are raised together. If the nonshared environment is important, concordance rates should be low for both monozygotic and dizygotic twins.

Actual heritability estimates for homosexuality in men vary between 0.25 and 0.65 (Mustanski, Chivers, and Bailey 2002; Långström et al. 2010), while the estimates are substantially lower in women. Heritability studies also suggest that the shared environment explains very little of the observed variation in the sexual orientation of men. The fact that little variation in sexual orientation is explained by variation in the shared environment has some obvious consequences for the mutability of homosexuality at the population level. Changing the shared environment is often easier than altering genes and nonshared environmental factors, if only because the nonshared environmental category functions as a sort of dumpster (as a science blogger once called it)[6] for every influence we cannot really identify as genetic or family-related.

That said, we should not read too much into the results of such heritability studies. Eric Turkheimer (2000), one of the world's most respected behavioral geneticists, has argued that results like the ones obtained for homosexuality similarly obtain for almost all human behaviors. It is usually the case that genes seem to be roughly as important as environmental factors in explaining behavioral variation, and that the nonshared environment often explains much more of the observed variation than the shared environment. Moreover, Turkheimer warns us that we should not interpret these results as evidence that nature generally trumps nurture. The results definitely do show that genes influence sexual orientation, but they do not show that their influence is bigger than the influence of the environment. This is one important takeaway from the heritability studies on homosexuality.

For a number of reasons, it is probably unwise to draw any stronger or more precise conclusions from these studies. First, heritability studies of sexual orientation tend to have low statistical power because it is often difficult to find sufficient pairs of gay twins in the databases that are used for heritability studies. When statistical power is low, one of the consequences is that only exceptionally and spuriously large effects will be reliably detected by the study, thus leading to an overestimation of the effect (Button et al. 2013). Therefore, it is likely that the true magnitude of the genetic contributions to the variation in sexual orientation is smaller than reported. Second, because many twin studies, including Kallmann's (1952a, 1952b), are family studies involving twins who were raised in the same homes, it is hard to disentangle the contribution of genetics and the environment. It is usually assumed that the environments of monozygotic and dizygotic twins are equal, but it is quite possible that parents treat monozygotic and dizygotic twins differently,[7] which could lead to higher concordance rates for monozygotic twins (Stein 1999). In principle, this issue could be resolved by so-called adoption studies, which study monozygotic twins reared in different environments, but such samples are hard to find (but see Eckert et al. 1986).

However, from the perspective of this chapter, the most important problem with a population measure like heritability is that it tells us very little about individuals.[8] Thus, the heritability of homosexuality cannot really tell us to what extent genes determine an individual's sexual orientation, nor to what extent orientation can be changed by environmental interventions. Because heritability is a property of the population, it only captures the fixity (and hence the mutability) of homosexuality in a population, and it does not tell us much about its developmental fixity. Therefore, it cannot say to what extent someone's homosexuality is immutable or irreversible.

All in all, it is fair to state that heritability would be a rather poor measure of innateness (Mameli and Bateson 2011). Measuring the heritability of homosexuality only provides a rough estimate of how much of the variation in the population can be changed by tweaking the environment (within the existing environmental parameters), with estimates ranging between 25 percent and 65 percent. More importantly, heritability is definitely not a measure of one of the main features of innateness, that is, developmental fixity, because it is purely a population measure.

Innateness as Canalization

When laypeople claim that an attribute or behavior of an organism is innate, what they sometimes mean to say is that one should look for the causes of that trait not in the organism's environment but solely in its genes. However, this meaning of the word "innate" has no scientific value at all. For if one claims that the environment has no causal influence whatsoever, it is then implied that the property would also exist without any environment, which is clearly absurd. While one might wonder if George Michael would still have been gay if he had grown up in a different environment, one cannot reasonably ask whether his sexual orientation would have changed if he had not grown up in *any* environment. The question is meaningless because no one, not even George Michael, can develop outside of an environment.

Faced with this objection, one could rephrase the innateness question as asking whether homosexuality is determined *primarily* by nonenvironmental factors, such as genes or hormones. Perhaps it does make sense to ask whether George Michael's sexual orientation is innate if less than 20 percent of its causes were environmental. However, this line of inquiry does not offer much relief either. After all, how should one determine the extent to which a person's sexual orientation is due to environmental factors? That question is as meaningless as asking which part of a drum is most responsible for its sound: the drumhead or the drumstick? (Block and Dworkin 1976; Lewontin 1982). In other words, it is nonsensical to claim that the sexual orientation of an individual is 10, 50, or 80 percent genetic. This way of looking at developmental fixity is ill fated.

Another and more promising approach is André Ariew's proposal to define innateness in terms of "developmental canalization" (Ariew 1996). For those familiar with the history of developmental biology, Ariew's proposal is not entirely new. The geneticist and developmental biologist Conrad Waddington coined the term "canalization" in 1942, with the explicit purpose to capture developmental fixity in a scientifically fruitful way. Building on Waddington's work, Ariew (1999, 117) defines innateness as canalization and canalization as "the degree to which the developmental process [of a trait] is bound to produce a particular end state despite environmental fluctuations both in the development's initial state and during the course of development." For example, even in widely different environments, most adolescents grow pubic hair. However, some adolescents don't, including those who have an intersex condition like complete androgen insensitivity syndrome. In Ariew's view, pubic hair is an innate characteristic in most humans, and the absence of pubic hair is innate in some intersex people.

If innateness should be primarily concerned with developmental fixity, then we agree with Ariew that Waddington's concept of canalization is a good candidate to capture such fixity. One relatively minor problem with Ariew's proposal is that developmental fixity does not always entail overall fixity: a trait can be strongly canalized because environmental perturbations are unable to alter its development, but the same trait may nevertheless be reversible in that it can be changed after it has developed. Within the ethically fraught debate on homosexuality and innateness, this issue is highly relevant. Proponents of so-called conversion therapy, for example, maintain that they do in fact help people become heterosexual, while recognizing, perhaps grudgingly, that they cannot prevent homosexuality from emerging altogether (Spitzer 2003).

A second and larger problem is that even if strong canalization were the best possible way to conceptualize the innateness of homosexuality, there are no available studies that directly measure such canalization. Moreover, it is highly unlikely that such studies will become available in the future because, as Turkheimer (2000, 161) notes, "ethical considerations prevent us from bringing most human developmental processes under effective experimental control." That said, there is a lot of biological research that seeks to identify biological causal factors in the development of homosexuality, and some of this research bears indirectly on the issue of canalization. Maybe somewhat surprisingly, however, such research mostly suggests that homosexuality is *not* strongly canalized, in at least one interpretation of the term "canalization." Here we give two examples of studies that provide indirect evidence that homosexuality is not strongly canalized.

First, recent genome-wide association studies (GWAS) on homosexuality, including one study mentioned earlier (Ganna et al. 2019), have identified a number of genes that are likely to affect the probability of the trait's development (Sanders et al. 2017). In the context of homosexuality, as in so many other contexts, genes fulfill their regular role as a probabilistic cause that increases the odds for a given individual to develop a homosexual orientation. It is perfectly possible, however, for a man to have all these identified genes and still be heterosexual. GWAS provide evidence against the strong canalization of homosexuality because they suggest that the environment plays at least some part in determining whether a given genotype develops into a homosexual or a heterosexual phenotype. Some genotypes are only a bit more likely to result in homosexual phenotypes, but as far as science currently knows, there is no genotype that results in a homosexual orientation in all natural environments of our species. The relatively low concordance of homosexuality for identical twins underscores this point (Gavrilets, Friberg, and Rice 2018).

A second line of research questioning the strong canalization of homosexuality focuses on birth order and intrauterine hormonal exposure. One of the few robust findings in scientific research on male homosexuality is the so-called fraternal birth order effect. In a series of papers, Ray Blanchard and his colleagues demonstrate that for every older brother, a boy's chances of becoming (or being) gay increase by 33 percent (see, e.g., Blanchard 2004, and further discussion of this effect in chapter 3). Given the odds that a male only child (or firstborn) is homosexual of approximately 2 percent, a boy with two older brothers has a 3.5 percent chance to become gay, and a boy with four older brothers a 6 percent chance. According to Blanchard, the origins of this birth order effect are to be found in the immune system of pregnant women. His hypothesis is that male homosexuality is at least in part the result of a mother's immune response to antigens produced by successive male fetuses—a reaction that would occur during the first months of pregnancy and disrupt the sexual differentiation of the fetal brain. Consequently, later-born sons would have a slightly more "feminine" brain than their older brothers. Blanchard further speculates that this would cause them to be more sexually attracted to men. Recent research suggests Blanchard may be right, as it shows that mothers produce specific antibodies against a Y-linked protein—antibodies that are indeed associated with a homosexual orientation in male offspring (Bogaert et al. 2018; Balthazart 2018). This finding did not really come as a surprise, since it was already known that the birth order effect also held for brothers separated shortly after birth and raised in different families (Bogaert 2006). It seems, then, that the fraternal birth order effect has its origins in the womb.

If we include maternal wombs in our list of environments, and if Blanchard is right in claiming that different uterine environments cause differences in sexual orientation, then it seems fair to conclude that the fraternal birth order effect also questions the strong canalization of male homosexuality.

It is important to stress that all these studies focus on factors that are colloquially called "biological," such as genes, neurons, immune reactions, and hormones, yet they can also be used to argue *against* the innateness (as canalization) of homosexuality. The reason is that these biological factors include the fetus's biological reaction to a particular environment, that is, its mother's womb. If Blanchard is right, environmental fluctuations do impact the development of sexual orientation, which implies, at least according to Ariew's definition of canalization, that homosexuality is not strongly canalized (or innate).

Of course, one could consider Ariew's definition too strict. After all, his definition is about the developmental fixity of *genotypes* and thereby excludes other types of developmental fixity. Suppose, for instance, that maternal stress or diet completely determines sexual orientation in the second trimester of pregnancy. For Ariew, this finding would imply that homosexuality is not canalized at all. However, it would also imply that sexual orientation is immutable after the second trimester of the fetus. For many, homosexuality would then appear to be developmentally fixed. Such looser interpretation of innateness (and canalization)[9] as developmental fixity would make it somewhat easier to determine the innateness of homosexuality. Still, it is rather unlikely that sexual orientation is entirely determined by maternal diet and stress levels during pregnancy. Many researchers believe homosexuality to be a complex trait that emerges out of very complex processes that often interact with each other (see, however, LeVay 2011). Therefore, it is difficult to determine at what phase of development sexuality becomes hard to change.

In our view, the research results we discussed do not exclude the possibility that particular social environments are necessary for the biological factors to have an effect on sexuality. As a prominent researcher of the biology of homosexuality admits, our current biological knowledge does not rule out the possibility that "all the biological factors . . . only produce a predisposition to become homosexual, and [that] these predispositions can only develop in a specific set of psychosocial contexts that are not yet identified" (Balthazart 2011b, 159). (Kallmann would probably smile his approval.)

To be clear, we are not denying that sexual orientation in adults and adolescents is really, *really* hard to change and that it becomes quite fixed at some point in the individual's development. In that sense, it clearly is somewhat canalized. We disagree, both on evidential and on moral grounds, with supporters of conversion therapy, who believe that it is possible to "cure" adolescent and adult homosexuals from their sexual orientation. In the past, conversion therapies have come in many forms. For a long time, aversion therapy via electric shocks was one of the more common methods. Male "patients" were shown photos of both men and women, with and without clothes, and given an electric shock when aroused by photos of naked men. The experimenters hoped that the "patient" would continue to avoid such images over time and develop an interest in photos of (naked) women instead (G. Smith, Bartlett, and King 2004). In some older studies, researchers used apomorphine, a substance that causes severe nausea, instead of electric shocks. Psychoanalysts also offered their services, which were very popular in the United States from the 1950s to the 1970s. From the 1980s onward, conversion therapy came under increasing attack, partly because they supported the disorder view of homosexuality and partly because they were not as effective as their practitioners believed (if they were effective at all). Worse, and perhaps unsurprisingly, qualitative research shows that conversion therapies can and do have seriously harmful effects on "patients" (Drescher et al. 2016). Still, even at the beginning of the twenty-first century, there were bona fide psychiatrists who believed that conversion therapy could indeed deliver its intended result (Spitzer 2003).

We do worry, however, about the assumed connection between immutability and innateness, particularly in the context of homosexuality. Some immutable traits are evidently acquired. For example, we, the authors of this book, speak Dutch, which is a capacity we learned somewhere between the ages of one and three. Having learned this, it would be a challenge to try and lose it again, at least without injuring our brains' speech centers. The immutability or developmental fixity of our capacity to speak Dutch thus does not imply its being innate. Likewise, even though homosexuality is not easy to change after a certain critical period in human development, this is clearly not enough to call it innate.[10]

In other words, if scholars like Hirschfeld believe homosexuality to be innate because it is hard to change, they likely commit a fallacy called "affirming the consequent" ("if a, then b; b, therefore a").[11] It may well be true that most innate traits are immutable, but not all immutable traits are innate.

WHY WE SHOULD AVOID INNATENESS CLAIMS

Behavioral genetics and developmental biology are well-respected sciences. Although there are all sorts of difficulties in determining the heritability and canalization of homosexuality, and although such determination would require ethically questionable methods, including experiments on human fetuses, it is in principle possible to do so. However, given the current state of biological research on human male homosexuality, we can only conclude that homosexuality is at least somewhat heritable and somewhat canalized. These are cautious but informative claims, and they are justified by the evidence that scientists have gathered over the past couple of decades.

That said, we do not think it advisable to replace heritability and canalization claims with innateness claims, even if they would be similarly nuanced ("homosexuality is probably somewhat innate"). We base our advice on three considerations.

First, there are convincing counterexamples to all existing accounts of innateness, including heritability and canalization accounts. Earlier on, we mentioned the so-called minimal condition, which stipulates that any innateness account should avoid conceptualizing traits as both innate and learned. Still, such traits do exist. Philosopher Richard Samuels (2002) gives as an example the belief that water is wet. We have learned that belief, but it is a belief that every human being shares regardless of the environment in which they grew up, hence it is strongly canalized for our genotypes. Similarly, some learned behaviors have a heritability of 100 percent. If all individuals of a population share the same environment, all variation in that population in literacy, for example, will be completely due to genes, even though every single word these individuals use is learned.

Second, we should avoid innateness claims because they will inevitably be less precise than similar relevant claims, including heritability and canalization claims. Unlike the concept of innateness, the concepts of heritability and canalization have precise meanings in the sciences, so they cause less confusion. Moreover, if all three concepts had meanings that were equally precise, there would be little reason to prefer innateness to heritability or canalization. We would lose no information by using any of the latter two concepts.

Third, innateness claims about homosexuality can and do lead to faulty inferences. In a series of studies, philosophers Paul Griffiths, Edouard Machery, and Stefan Linquist (2009) have experimentally shown that innateness suggests fixity, typicality, and teleology all at once, and that it is difficult to disentangle these three aspects. For example, if a developmentally fixed trait such as the song of a corvid species is said to be innate, people, including biologically trained scientists, will usually interpret it as an *adaptive* trait. Still, many adaptations, such as heart rate, are flexible, whereas many developmentally fixed traits, such as cystic fibrosis, are maladaptive. Of course, one can always warn people that they are likely to misinterpret claims of innateness. But then why use the term in the first place, given that there are scientifically valid alternatives that avoid such misinterpretations? Again, we do not want to claim that homosexuality is not innate. We are rather emphasizing that such a denial would be as confused as its antithesis, that is, the confirmation of homosexuality's innateness.

Nativists about homosexuality might counter that while this is all true and innateness is indeed an "irretrievably confused" concept (Griffiths 2002), their nativism should be understood as a denial of the so-called constructivist or constructionist view. And, indeed, the prime target of many nativists is the view that homosexuality is a social construction. However, in the next section we argue that because there are so many kinds of social constructivism, anticonstructivist nativism can only be taken seriously on the condition that it clarifies first what kind of social constructivism it rejects.

SOCIALLY WHAT?

Within the social sciences, nativism is sometimes seen as the opposite of social constructivism (Bickerton 1997; Peterson 2012). When nativists claim, for example, that homosexuality is innate, they often mean that it is not some kind of social construction. To make sense of such claims, however, we first need to define social construction.

Constructivists claim that homosexuality was socially manufactured in nineteenth-century Western Europe, a period that birthed a new way of talking and thinking about homosexual behaviors and desires. Many actors were involved in shaping this new mode of thought, but its clearest expression can probably be found in psychiatric theorizing at the time. Although social constructivists admit that homosexual behaviors and desires have been observed in many cultures and eras across the world, they argue that the modern, Western way of thinking about homosexuality was new and unique. (In chapter 3, we discuss this claim in more detail.) Constructivists also argue that this new way of thinking produced a new form of homosexuality—*modern homosexuality*. Some constructionists even go as far as to claim that there was no such thing as homosexuality in most premodern or non-Western societies. In *One Hundred Years of Homosexuality*, for example, historian David Halperin (1990, 29) states: "Although there have been, in many different times and places (including classical Greece), persons who sought sexual contact with other persons of the same sex as themselves, it is only within the last hundred years or so that such persons (or some portion of them, at any rate) have been homosexuals."

This brief summary of constructivist thinking about homosexuality, however, obfuscates the many variations in contemporary social constructivism. There may be some consensus among constructivists about the locus of construction (*where* did it happen?), but there is much less about its subjects (*who* did the construction work?), its objects (*what* was constructed?), and its rationale (*why* was homosexuality constructed at all?). Meanwhile, philosophers have produced a range of answers to questions about the nature of the construction work (what does it *mean* to construct?), distinguishing among different levels of construction that are relevant vis-à-vis the debate with nativists.

Some constructivists view the meaning of the term "homosexuality" as a construction (semantic constructivism); others consider the category or concept of homosexuality to be constructed (epistemological constructivism); still others think that the phenomena described by these terms or categories are constructed (ontological constructivism) (Mallon and Stich 2000; De Block and Lemeire 2015).[12] In the first part of this section, we explore these three variants of social constructivism and briefly sketch if, and how, they conflict with nativist views about homosexuality. In the second part, we consider whether it makes sense to define nativism as anticonstructivism.

Constructing Homosexuality

Semantic constructivists hold that there was no so-called nominal essence of homosexuality in premodern and non-Western societies. Semantically, the term "homosexuality" does not refer to anything before the nineteenth century. A social constructivist may argue, for example, that "homosexual desire" only refers to the desire for homosexual activities, as displayed by individuals who attribute their desire to their identity, whereby their identity is partially formed by psychiatry. If this is the correct semantic meaning of the term "homosexual desire," then it clearly cannot refer to any desire or mental state that precedes the advent of psychiatry. Consequently, it makes sense to say that homosexuality did not exist in ancient Greece.

The question is, however, whether this semantic variant of constructivism is an interesting one. Very few people, even among nativists, would deny that contemporary psychiatry originated in nineteenth-century Western Europe and that it had some effect on the self-identification of those subjected to its practice. On this level, therefore, there is no real disagreement between nativists and constructivists, except perhaps about the right theory of meaning and reference (Mallon 2016). If you think that the meaning of "homosexuality" is determined by a thick description that includes the effects of nineteenth-century psychiatric practices and theories, then the term cannot refer to anything that existed in premodern and non-Western societies. If, by contrast, you subscribe to a thin description of the theory of meaning, then particular behaviors and desires in ancient Greece can count as homosexual behaviors and desires, even though they may well differ in notable ways from modern homosexual behaviors and desires. (The reader will remember that we opted for such a thin definition of homosexuality in our general introduction.) Many nativists believe that homosexuality is a cross-cultural universal, and thus they adhere to a thin description of the theory of meaning—a preference that, in our view, does not affect their scientific methods or arguments.

Epistemological or categorical constructivism, the second variant of constructivism, argues that various nineteenth-century social forces caused homosexuals and heterosexuals to be classified as two distinct kinds of people. As these forces did not exist before the nineteenth century, they produced something truly new, that is, categories of people whose identity was largely determined by their sexual behaviors and desires. Other cultures and eras organized their sexual landscapes differently, by means of different categories. For example, in ancient Greece, popular sexual categories included the *eromenoi*, boys and male adolescents, and the *erastai*, married adult men. Despite their differences in age and social status, eromenoi and erastai maintained a close relationship in which sex played an important part. Both roles came with all sorts of requirements (Halperin 1990; Murray 2000; see also chapter 3). Eromenoi were supposed to embody youthfulness, for example, and to play a passive role in sexual activities; erastai were expected to marry and have children, while also focusing on their relationship with the eromenoi. Twenty-first-century Westerners do not use these categories anymore, which means we have a different understanding of homosexuality (Murray 2000). In fact, we think of the ancient Greek sexual landscape as positively weird, as Halperin (1990, 2–3) remarks:

> Everyone who reads an ancient Greek text, and certainly anyone who studies ancient Greek culture, quickly realizes that the ancient Greeks

were quite weird, by our standards, when it came to sex.... Like most of my scholarly colleagues in the field of classics, I ... didn't have a language for articulating systematically the discontinuities between ancient Greek sexual attitudes or practices and my own. Or, at least, I didn't have such a language until the mid-1980s when it was provided me by the new social history and, specifically, by the work of social constructionist historians of homosexuality, notably George Chauncey. What I, and many others, have learned from this work is that it is not the Greeks who were weird about sex but rather that it is we today, particularly men and women of the professional classes, who have a culturally and historically unique organization of sexual and social life and, therefore, have difficulty understanding the sex/gender systems of other cultures. One of the most distinctive features of the current regime under which we live is the prominence of heterosexuality and homosexuality as central, organizing categories of thought, behavior, and erotic subjectivity.[13]

Such differences in categorization can also have an impact on how we conceive of the origin or development of homosexuality. For example, according to our modern categorization, people no longer *become* homosexual but rather discover that they have been homosexual all along. Epistemological constructivists note that we started to think very differently about homosexuality in the nineteenth century—a new way of thinking that was based on new categories and new distinctions and dichotomies.

Whether these new categories of homosexuality and heterosexuality really carve our species' sexual nature at its joints falls, strictly speaking, outside of the purview of epistemological social constructivism. After all, demonstrating that social factors play a role in how we think about a particular phenomenon, or even that social factors determine whether we see something as a distinct phenomenon, does not mean that a phenomenon is not real or distinct. For example, capitalism's focus on productivity may have led to our modern way of thinking about sexuality in general, as Foucault argues, but perhaps capitalism was only necessary for us to understand what sexuality is and how to categorize it. Of course, constructivist histories about the causes of our new classifications explain our current use of such classifications, as well as many of our beliefs. As such, constructivists may undermine some nativists who claim that we have always used these categories and that these categories are human universals.

The third variant of constructivism, ontological (or causal) constructivism, goes one step further than its epistemological counterpart by holding not only that particular nineteenth-century social forces created new sexual categories and classifications but also that these novelties, in their turn, brought into existence all sorts of properties that are deemed typical of modern homosexuality. Some ontological constructivists even claim that these new categories and classifications produced an entirely new social role: the homosexual. Foucault, for example, famously contrasted modern homosexuality with premodern "sodomy" (Foucault 1978, 43). Many contemporary sex historians, including Halperin, have followed in Foucault's footsteps, although they also tend to disavow his disciplinary rigor in distinguishing among modern and other variants of homosexuality and pay more attention to historical continuities (Halperin 2002; Sedgwick 1990).

Unsurprisingly, ontological constructivism is the prime target of nativists, as it implies that homosexuality is a recent and entirely new phenomenon brought about by a society's sexual categories and classifications, rather than by an individual's biological make-up. In addition, there is the not-so-subtle constructivist hint that our current categories may not be carving nature at the joints. The very fact that a society would be able to produce new kinds of people simply by constructing and imposing new conceptual categories suggests that human nature is much more malleable than nativists tend to think. Many nativists thus take offense at constructivist thinking. In a paper on the psychobiology of homosexuality, the British psychologists Qazi Rahman and Glenn Wilson (2003, 1338) provide an example of a nativist rebuttal: "Social constructionists argue that sexuality is a fluid and dynamic property that can only be understood by the analysis of sociopolitical contexts, linguistic and narrative 'scripts' . . . [but] [r]esearchers within the field of sex research have . . . rebuffed social constructionism as a poor intellectual framework for understanding sexual orientation." In their view, the life sciences undercut the constructivist position, particularly by charting and explaining the biological origins of homosexual behaviors and desires (see also Gangestad, Bailey, and Martin 2000). But do they, really?

How Confused Is (Anti-)Constructivism?

The problem with this form of nativist criticism is that it does not show much interest in the sophistications of contemporary social constructivism. On the one hand, nativists could easily blame their opponents for this inaccuracy, if only because constructivists themselves often appear confused about what exactly they are saying and lump together claims that should be carefully distinguished.[14]

On the other hand, nativists should also acknowledge blame because they, too, rarely engage with specific constructivist claims and arguments. Rahman and Wilson (2003, 1338), for example, lazily dismiss social constructivism as an "incoherent body of postmodernist concepts emphasizing the subjectivity of scientific inquiry and method, and the relative nature and equal validity of conflicting epistemologies." Of course, they do argue, explicitly and in great detail, for the relevance of biological and psychobiological factors in studying the etiology of homosexuality, but rather than establishing the irrelevance of social factors in this context, they simply assume such irrelevance follows from their findings. However, this assumption is incorrect. In chapter 3 of this book, we explore a number of contemporary hypotheses about the evolution of homosexuality, in which biological (or "nativist") and cultural (or "constructivist") explanations are mutually compatible.

It seems, then, that nativists are misguided in targeting constructivist ideas about homosexuality. Constructivists, in their turn, rarely target nativist ideas, but they do become enraged by essentialism. When discussing *One Hundred Years of Homosexuality* in a later book, Halperin (2002, 10) notes that its "overriding purpose ... was to win the once-vehement debate between essentialists and constructionists over the constitution of sexual identity." Essentialists maintain that there is such a thing as the essence of homosexuality—an essential property that all and only homosexuals have and have always had. This property would allow us to distinguish neatly between homosexuality and heterosexuality, while at the same time ensuring an important homogeneity within the population of homosexual males. (See chapter 3 for more about essentialism.) In philosophical parlance, essentialists maintain that homosexuality is a natural kind. As such, one could compare sexual orientations to chemical elements, where essential properties take the form of atomic numbers.

Constructivists, by contrast, take the view that there are in fact many homosexualities (Murray 2000), or many ways of being homosexual, just like there are many ways of being heterosexual (Blank 2012), each of which derives from its specific social and cultural environment. And even though there is in today's Western societies a clear and rather sharp difference between homosexuality and heterosexuality, as well as a substantial uniformity within each of these categories, we must not forget, constructivists claim, that these features are produced and maintained by social rather than biological causes. Moreover, it would also be a mistake to assume that the emergence of modern homosexuality completely eradicated other forms of homosexuality, even in modern Europe (Sedgwick 1990). Of course, the debate between essentialists and constructivists often touches on the role that biological factors play in the etiology and development of homosexuality. After all, if biological factors suffice to explain the differences between homosexuals and heterosexuals, and if they suffice to explain most of the similarities among homosexuals, then the essentialist position would indeed be strengthened.

However, there are three reasons why biological data are rarely decisive in this debate. First, we already mentioned that social constructivist explanations are compatible with biological explanations. For example, one can argue that homosexuality is a social construction but that there are biological factors that facilitated the construction of this new reality. In fact, this explanatory strategy is often followed with regard to the ontology of race. While most philosophers and social scientists are convinced that race is a social construct and argue that the sharp differences we observe between, say, Whites and Blacks are produced by social forces, these theorists also acknowledge that our categorizations build on rather shallow but nonetheless genuine biological differences, such as differences in skin color. In chapter 3 of this book, we will further explore this compatibility in the context of evolutionary explanations of human homosexuality.

Second, essentialism does not need to be nativist, because it can emphasize that environmental factors cause the essence of homosexuality. The Freudian explanation of homosexuality, for example, is deeply essentialist, but it is also environmentalist (Stein 1999). According to Freud, homosexuals share many attitudes, fears, and desires because their fathers were too absent and their mothers too present during a critical period in early development. This psychodynamic explanation is both environmentalist and essentialist.

Third, even self-proclaimed anticonstructivists often seem to deny—at least implicitly—that there is such a thing as the essence of homosexuality. As mentioned before, in order for there to be any real essence of homosexuality, we must be able to find a property, or a set of properties, that is sufficient for universally determining who is and who is not homosexual. The existence of such a property or set of properties seems at odds with the very complex biological etiology of homosexuality, as described and analyzed by anticonstructivists such as Rahman and Wilson.[15]

For now, we want to point out that the debate between constructivists and nativists is likely as mixed-up as the innateness debate. While constructivist and anticonstructivist discussion can sometimes be illuminating, it is often confused and confusing. Both constructivists and nativists tend to misunderstand not only the position they are arguing against but also their own position in the debate.[16] More importantly, constructivism really aims not at nativism but at essentialism. Therefore, it is strange that nativists would define themselves as anticonstructivists, especially given the fact that most nativists do not embrace essentialism. Hence, rather than relying on the umbrella terms of "nativism" and "constructivism," it is often better to concentrate on the methods and data that underlie these terms.

CONCLUSION

Is homosexuality innate? We argue that the question is likely meaningless, and almost certainly "irretrievably confused" (Griffiths 2002), because the word "innate" has so many different meanings in so many different contexts. We could try to subsume some of those meanings under respectable scientific concepts, such as canalization and heritability, but there is no scientific concept that unites all the meanings of "innateness." And even if we let this go and assumed that, for example, "highly canalized" is a valuable substitute for "innate," we would still have to conclude that scientific progress has not sufficiently provided a conclusive answer to our question. The same is true for attempts to recast the innateness debate as a debate about the social construction of homosexuality. Science matters for some but not all the issues that surround the constructivist debate, and its participants often misunderstand the debate's central issues.

This conclusion may be disappointing, but it obviously does not mean that we are as ignorant about the causes and forms of homosexuality as the author of the *Problemata* was. In the past hundred years, our knowledge about endocrinology, genetics, developmental biology and psychology, social psychology, sociology, anthropology, and the history of human male homosexuality has increased exponentially. This increasing knowledge has created room for nuance and for interesting insights. It has also created a fertile soil for philosophical analysis, exploration, and speculation, to which we hope the following chapters will attest.

[CHAPTER 2]

Sham Matings and Other Shenanigans

On Animal Homosexuality

Natura maxime miranda in minimis.

JOHANN CHRISTIAN FABRICIUS

Birds do it, bees do it, even educated fleas do it.

COLE PORTER

Is there such a thing as animal homosexuality? And why should we care? These two questions are at the heart of this chapter. For the past two hundred years, biologists have been in two minds about this "love that dare not squeak its name," as a witty commentator recently described it (D. Smith 2004). One camp is skeptical, explaining away animal homosexual behaviors, for instance, as mere reflex phenomena that result from various circumstances, such as sex segregation or a failure in sex recognition. Surely, these behaviors should not be associated with homosexual desires, preferences, or identities, as they usually are in humans. Because of these and other differences, some biologists argue that the term "homosexuality" should be reserved for humans only.

Other biologists have been much less reticent and object to these skeptical views on both conceptual and empirical grounds. First, they argue that there are various definitions of concepts like desire and preference, some of which allow us to ascribe homosexual desires and preferences to some nonhuman animals. Second, they provide empirical evidence that suggests the existence of many dimensions of homosexuality, in addition to behaviors, including evidence that some male animals do prefer homosexual to heterosexual activities.

Despite their differences, both camps agree that there is a growing biological literature documenting males having sex with males in hundreds of species from all branches of the animal kingdom, both in captivity and in the wild. Apart from the question what it is exactly that they are having (when they are having sex), there are further questions to be answered: Why should we care? What is there to learn from studying animal homosexuality? There are many possible answers to these questions, and most of them relate to our interest in *human* homosexuality—its moral status, its nature, and its many causes. The eighteenth-century naturalist Georges-Louis Leclerc de Buffon (1954, 317) once said that human nature would be even more puzzling if animals did not exist, and the same can likely be said of homosexuality.

This chapter details a number of philosophical questions surrounding the connection between animal and human homosexuality. We argue, among many other things, that it need not be a category mistake to ascribe homosexual desires, preferences, or even orientations to some animals; that debates about the concept of homosexuality are often troubled by arguments about the more fundamental concept of sexuality; that zoologists are not necessarily guilty of anthropomorphism when using the term "homosexuality"; and, finally, that animal models can and do play an important role in research on human homosexuality.

We begin, however, with a rather detailed reconstruction of one of the very first scientific discussions about animal homosexuality: the nineteenth-century debate about homosexual copulation in cockchafers (*Melolontha vulgaris*). The *Melolontha* story reveals a surprising variety of early expert opinions about many aspects of homosexual behavior in animals, including its naturalness and innateness, its diversity, its mechanisms and determinants, its relationship with homosexual desire and preference, its moral implications, and, of course, its relationship with human homosexuality. As such, the story introduces us to most of the issues, both scientific and philosophical, that this chapter deals with. In addition, the story will reverberate in later chapters, when we deal with questions about the nature, the causes, and the morality of human homosexuality.

THE *MELOLONTHA* STORY

On the evening of Tuesday, June 2, 1863, in a large lemon grove outside the coastal village of Menton, the French entomologist Alexandre Peragallo made an extraordinary discovery. While witnessing the courtship displays of male fireflies (*Luciola lusitanica*), he noticed a number of *Luciola* males were being covered by male soldier beetles (*Rhagonycha melanura*). After barely recovering from this "gruesome" observation, he asphyxiated three of the couples with sulfur to document the "monstrosity" for his colleagues at the Société Entomologique de France (Peragallo 1863, 661). Peragallo

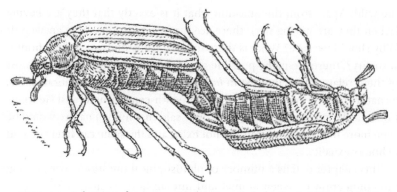

Figure 2.1: A drawing of same-sex coupling in cockchafers by A. Clément (1896). Source: Gadeau de Kerville 1896b, 85.

believed the coitus to be "fully complete, certain, and even intentional," and so he concluded that both species, in the end, were equally guilty, although in different ways: "As I am absolutely confident of the sex of both insects, and that they are both male, I cannot but assent to a blatant immorality on the part of *Rhagonycha*, and a guilty complicity on the part of the male *Luciola*" (663).[1]

Peragallo's astonishment is slightly surprising, since he wasn't the first to document homosexual mating in the insect world, even within the close confines of his own society and its specialist journal, the *Annales de la Société Entomologique de France*. Four years earlier, in 1859, the very same journal had already published a brief report on the coupling of two male common cockchafers (*Melolontha vulgaris*). The author of the report, Alexandre Laboulbène, received the specimens from a colleague who had caught them a few years earlier in Normandy. The colleague found it hard to believe that both insects were actually male, but Laboulbène's meticulous dissection proved that they were. It also revealed some interesting technical details about male-male *Melolontha* copulation. From the outside, male-male couplings didn't seem to differ much from male-female couplings, wherein males dangle backward off the female's back, seemingly asleep, while their penis and anal stylet provide the sole point of attachment between both partners (Laboulbène 1859, 569). Figure 2.1 is a beautiful late nineteenth-century pen drawing of a male-male cockchafer couple in action—a drawing that, according to one commentator, is also "the first known pictorial representation of non-human, same-sex coupling" (Brooks 2009, 153).

In male-male couplings, the penetrating male inserts his penis into the external reproductive orifice of his partner. Consequently, the latter's genitalia are pushed back into his abdomen. Unlike Peragallo, Laboulbène did not express any moral concerns regarding the insects' behavior. Nevertheless, he, too, concluded his note with a hint of surprise: "This mating behavior is very unusual and peculiar, and I don't know of any other genuine example" (Laboulbène 1859, 570).

Apparently, Laboulbène had not seen the very first report of male-male copulation in an insect species, an 1834 publication in the German interdisciplinary science journal *Isis von Oken* (Kelch 1834). Here a German schoolteacher, August Kelch, vividly described his disbelief upon discovering in a forest near Ratibor (currently Racibórz in Poland) a male common cockchafer covering a smaller male forest cockchafer (*Melolontha hippocastani*). Like Laboulbène's colleague, Kelch initially believed the penetrated male to be a female with male antennae.[2] Kelch changed his mind only when he discovered the penetrated male's genitalia hidden in his rear cavity, exactly where Laboulbène would find them fifteen years later. Concluding his observations, Kelch (1834, 737–38) attempted to explain the insects' behavior: "So it was clear that the male *Melolontha vulgaris*, being the bigger and stronger one, had simply overpowered the smaller and weaker male forest cockchafer by exhausting it; and only because of its dominance, was it able to force itself onto the other, so to speak."

So why would these beetles engage in homosexual behavior? During the second half of the nineteenth century, two conflicting sets of explanations arose in the entomological literature.

Some authors hypothesized that most male-male copulation in the insect world was due to the penetrating male's inability to properly identify the sex of his partner. In their view, some males simply mistook other males for females—an error often precipitated by circumstances, such as uncontrollable lust on the part of the penetrating partner or exhaustion on the part of the penetrated. The German entomologist E. Doebner (1850, 328), for example, suggested that male cockchafers "make use of other males to satisfy their violent procreative urges, probably taking them for, and overpowering them as, females in their blind passion."

In France, physician and neurologist Charles Féré was a vocal defender of the error hypothesis. Féré (1897, 496) argued that hot-blooded males are tricked into "unnatural acts" by the abundance of what are now known as female sex pheromones found on the bodies of other males that had recently copulated with females. Exhausted by their heterosexual mating, these males tend to be easy targets: "A male cockchafer who just copulated with a female is necessarily permeated by smells that are designed to attract another male, and besides he is half-dead with exhaustion and utterly incapable of offering resistance; he simply resigns himself, we should say, to the mistake" (498).

Féré conducted a series of pioneering scientific experiments to test this hypothesis. The main experiment tracked the number of male-female (x) and male-male (y) copulations in three different groups. A first group consisted solely of females and virgin males ("mâles neufs"), or males with no previous sexual experience whatsoever. The second group consisted of a mixture of females, virgin males, and males soaked with female pheromones ("mâles imprégnés"), while the third group contained a mixture of females, virgin males, and males that had recently copulated with a female ("mâles émérites") (Féré 1899, 79–85). Féré found that every copulation in the first group was heterosexual ($x=300$; $y=0$), while he counted only two homosexual copulations in the second group ($x=206$; $y=2$). In the third group, however, nearly 10 percent of all couplings were homosexual ($x=193$; $y=17$). The finding that most homosexual copulations occurred in the third group seemed to fit Féré's hypothesis. Unlike their "emeritus" colleagues in the third group, the "soaked" males in the second group had not exhausted themselves by mating with a female and were therefore able to escape or resist "blinded" virgin males. Féré's hypothesis was further supported by the finding that the large majority of penetrated males, in all groups, were revealed to be either "soaked" or "emeritus." In Féré's view, male-male cockchafer coupling was nothing more than "a trap" (82) in which both parties lose out.

Other scientists, however, drew on the very same cases and observations to cast doubt on the error hypothesis. In their view, homosexual copulation in cockchafers was due neither to coercion or violence nor to any kind of error. Rather, they claimed that some male cockchafers, in some circumstances, *desire* or even *prefer* to have sex with males over females, and they simply act or choose accordingly. These males do not mistake other males for females; they just want males.

One of the earliest champions of this hypothesis was the Russian diplomat and entomologist Carl Robert Osten-Sacken. In a brief paper in *Entomologische Zeitung*, he discussed and confirmed his colleagues' observations of male-male cockchafer coupling while at the same time reassessing their conclusions (Osten-Sacken 1879). He also presented his own work on the topic, which acknowledged, for example, traces of violence on the penetrated male's body. In one of his specimens, the penetrated male's penis was dangling by a thread from the rear of its body. Unlike Kelch, however, Osten-Sacken argued that unilateral violence could not be the end of the story. In his view, some of the technical details of male-male *Melolontha* copulation would be utterly inexplicable unless one assumed that both partners somehow cooperated voluntarily in the act. This assumption, in fact, provided the main argument for Peragallo's moral outrage, so it's no surprise that Osten-Sacken (1879, 118) alluded to Peragallo's 1863 report: "The salacity [*Sinnlichkeit*] must be coming from both parties; since merely by means of violence, and without the 'coupable complaisance' [guilty complicity] of the passive male, this act could not be accomplished mechanically."

Osten-Sacken reasoned that homosexual copulation in cockchafers might be associated with homosexual desire. In France, his colleague Henri Gadeau de Kerville went one step further by associating homosexual behavior with homosexual preference (see also Féray 2004). On the evening of February 26, 1896, Gadeau de Kerville caused quite a stir among the audience of the annual conference of the Société Entomologique de France. In a talk about "sexual perversion" in male beetles, he distinguished between two kinds of "pederasty" in the insect world: "pederasty by necessity" (*pédérastie par nécessité*) and "pederasty by preference" (*pédérastie par goût* or *par préférence*) (Gadeau de Kerville 1896b, 85).[3] The former involves male-male copulation in the absence of available females of the same species, while the latter refers to male-male coupling in the presence of such females.

In a follow-up paper published the same year, Gadeau de Kerville (1896a) described the debate that ensued at the conference and condensed the audience's responses into two basic objections. First, some scholars objected to the use of the word "pederasty" in his account since, etymologically, the term was specific to a kind of age-asymmetric sexual relationship between married men and (young) adolescents in antiquity. (In our general introduction and in the previous chapter, we described this kind of relationship as intergenerational or ritualized homosexuality; we discuss this subject further in chapter 3.) Lacking such asymmetry, his opponents argued, the presented cases of male-male *Melolontha* copulation could not be "real acts of pederasty" (Gadeau de Kerville 1896a, 1). Gadeau de Kerville justified his wording by claiming that the meaning of "pederasty" had changed since antiquity and that his colleagues, including the German sexologist Albert Moll, now used it to refer to a specific subgroup of homosexuals who practiced anal intercourse. (Moll indeed used the term "pederasty" rather frequently, in particular in the chapters on animal homosexuality in *Untersuchungen über die Libido sexualis* [*Libido sexualis: Studies in the Psychosexual Laws of Love*; Moll 1898, 368–75, 492–94.])[4] Therefore, cockchafers could be called pederasts to the extent that male-male cockchafer copulation involves anal penetration.

A second objection to Gadeau de Kerville's talk at the Société Entomologique de France involved his distinction between two different kinds of pederasty. While none of his colleagues doubted that the physical needs of males might sometimes urge them to copulate with other males when females are unavailable ("pederasty by necessity"), his concept of "pederasty by preference" was met with violent protest. The claim that some male cockchafers would somehow prefer to have sex with males rather than females was dismissed as implausible and even preposterous.

Determined to make his case, Gadeau de Kerville parried with two arguments. First, he presented a number of published case studies showing that male-male insect copulation did indeed occur despite the ready availability of females. Quoting extensively from Peragallo's report on fireflies and soldier beetles, for example, he noted that the latter must have had access not only to *Rhagonycha* females but also to female fireflies, yet they still had sex with *Luciola* males. "Therefore," he concluded, "these *Rhagonycha* males were pederasts by preference, not pederasts by necessity" (Gadeau de Kerville 1896a, 10). Second, he rebutted rival hypotheses. One such hypothesis, later taken up by Charles Féré, was that *Melolontha* males could perhaps be deceived into copulating with other males who still exuded female pheromones after having coupled with a female (7–8). Gadeau de Kerville considered this explanation implausible: "On reflection it is hardly plausible that a male who has coupled with a female, and then uncoupled, would give off a female odor that is more powerful than the odor of the females in his vicinity" (8). Gadeau de Kerville realized that these observations and arguments were not completely decisive evidence for the existence of "pederasty by preference" in insects, but he hoped, at least, to formulate a plausible hypothesis that deserved further study.

At the end of his paper, Gadeau de Kerville provided a third and final argument in support of his hypothesis—an argument from analogy. Pederasty by preference, he argued, does exist in the human species. Moreover, there were interesting similarities between male-male cockchafer copulation and human (male) homosexuality, including the technique of anal penetration (5). In fact, humans and other animals share many sexual practices, ranging from masturbation[5] to oral stimulation and anal penetration. Why would it then be silly to suggest that some animals prefer to have sex with same-sex partners? Most interestingly, Gadeau de Kerville was even willing to consider the hypothesis that some animals, like some human males, were born pederasts: "Why should we not accept the existence of insects that also have a congenital [or innate] homosexual preference . . . ?" (11; see also H. Ellis and Symonds 2008; Moll 1898).

Féré disputed most aspects of Gadeau de Kerville's analysis. While he did not deny that some animals display homosexual behavior, he thought it unlikely to assume that they would *prefer* such "unnatural acts" (Féré 1899, 73), let alone that such preference was innate. Animals with an innate homosexual preference, he argued, would simply be doomed to die out. They would not be motivated to mate with females, thus depriving themselves of the possibility to produce offspring. (This argument is also at the heart of some contemporary evolutionary research on human homosexuality, as we explain in the next chapter.) Moreover, Féré continued, innately homosexual animals would likely be mistreated by their congeners because of their sexual interests (85). Interestingly, Féré believed that homonegativity was not a uniquely human phenomenon. However, he did not provide any evidence for that claim. Féré concluded: "In actual fact, the existence of sexual inversion as we know it in humans—the inborn homosexual love—has not been established at all in animals: their perversions are accidental or acquired" (76).

The actors of the *Melolontha* story disagreed on everything, except perhaps on the depravity of their research subjects. Implicitly or explicitly, most nineteenth-century entomologists repudiated homosexuality. Even those who allowed for the possibility that some male cockchafers preferred homosexual to heterosexual copulation, and that some of them may be innately homosexual, certainly did not conclude that animal or human homosexuality was morally acceptable. On the contrary, they abhorred any kind of homosexuality, deplored its (apparently) rising prevalence, and described animal homosexuality as "morbid" (Féré 1899, 72), "monstrous" (*monstrueux*; Peragallo 1863, 661), "unnatural" (*widernatürlich*; Osten-Sacken 1879, 118), and even "hideous" (Gadeau de Kerville 1896a, 3).

It is rather ironic, then, that their findings and hypotheses would play an explicitly emancipatory role in the work of one of their contemporaries, the activist and former lawyer Karl Heinrich Ulrichs, whom we referred to in the introduction to this book. A homosexual himself, Ulrichs was one of the first advocates of equal rights for human homosexuals, or "Uranists" (*Urnings*), as he called them. His life's work argued that homosexual desires in both men and women were a natural feature of their personality, on par with heterosexual desires. As such, he claimed, human homosexuality should not be subject to legal, political, or social prohibitions. In his attempts to subvert such prohibitions, which were in effect in many European countries at the time, Ulrichs appealed to findings, concepts, and theories from various sciences, including forensic medicine, embryology, and zoology (Ulrichs 1994; Brooks 2010). He referred to Osten-Sacken's report on male-male *Melolontha* copulation as evidence of consensual homosexual coupling in animals (Ulrichs 1994, 684), and he anticipated Gadeau de Kerville's claim about the innateness of homosexual desires in beetles and other animals. In a pamphlet from 1870, Ulrichs (1994, 604–5) summarized his views: "The *Urning*, too, is a person. He, too, therefore, has inalienable rights. His sexual orientation is a right established by nature. Legislators have no right to veto nature; no right to persecute nature in the course of its work; no right to torture living creatures who are subject to those drives nature gave them.... The battle against nature is a hopeless one."

In Ulrichs's view, the existence of homosexual desires in animals provided proof of the naturalness of human homosexuality, and therefore of its normality. While the protagonists in the *Melolontha* story might have disagreed, Ulrichs claimed that their position entailed all sorts of absurdities, including the idea that insects could actually be mentally ill. In the final lines of his *Critische Pfeile* (Critical arrows), he notes: "Speaking for those 'born according to natural law' [i.e., human homosexuals] . . . is that occurrence of man-manly sexual acts among insects. My opponents must have excluded this phenomenon from the field of natural history. They had to add it to the School of Instruction in Diseases, Psychiatric Division, Department of Mentally Ill Insects" (Ulrichs 1994, 688).

In later chapters of this book, we will elaborate on the idea that animal homosexuality can help us normalize, or at least decriminalize, human homosexuality. A hundred years later, for example, in the 1970s, it would play an important role in gay activists' objections to the view that homosexuality is a mental disorder.

Ulrichs's line of argument did not impress many of his contemporaries. According to some commentators, he failed in his first book to provide good arguments for the claim that male homosexuality might be natural or innate, in seeking "to correct public opinion through all kinds of mostly irrelevant observations, quotations, anecdotes, outpourings of his heart" (anonymous reviewer [1865], quoted in Dworek 1990, 44). Another critic blamed Ulrichs for being vague about the supposed naturalness of homosexuality: "[The author] is at pains to furnish all kinds of things in order to show that 'nature has created pederastic lust.' What could he be thinking of with the word 'nature'?" (quoted in Dworek 1990, 44). It was only in 1879, in one of his later works, that Ulrichs defended against this criticism by referring to contemporary research on animal homosexuality.

Finally, most of Ulrichs's contemporaries emphasized that the kind of homosexual behaviors that are found in animals could not in any way serve to legitimize human homosexuality in the way that Ulrichs had hoped they might. Féré was one of them: "In animals, the isolated or collective anomalies of sexual satisfaction are nothing but reflex phenomena or automatisms. Their manifestations, though they may appear to be common, do not in any way imply that these anomalies would be legitimate in humans" (Féré 1899, 87).

Féré acknowledged that some humans are innately homosexual, but in his view, such homosexuality isn't natural at all, both because it lacks an analogous phenomenon in the animal world and because congenital homosexuals do not obey the natural law of reproduction: "If a species perpetuates itself, it is most certainly not due to inverts" (Féré 1896, 5). And even if congenital homosexuality was natural in some sense of the word, for example, because it also occurred in some animal species, it certainly wasn't to be considered biologically or medically normal. If anything, Féré argued, congenital homosexuality was a morbid, degenerate condition—a disease: "However superior he may be, the invert is always a degenerate" (5).

In the second half of the nineteenth century, most scientists believed that the vast majority of homosexuals suffered from psychiatric problems, and even those who managed to maintain themselves in society were seen as degenerates. Degenerationism was a very popular way of thinking in nineteenth-century medicine and psychiatry. It held that countless human problems and disorders were due to a downward, degenerative process that, once initiated, would cause a bloodline to deteriorate and eventually and inevitably end in complete idiocy. Homosexuality was one of the many stops on the way to that end, and so Féré advised homosexuals not to have children at all.[6] By contrast, those who, like animals, engaged in accidental homosexual behavior could and should be educated and treated in order to restore their humanity. Concluding his chapter on animal perversions in *L'instinct sexuel* (The sexual instinct), Féré noted that accidental homosexuality in humans represented nothing but a temporary resurgence of man's animal nature: "The symptoms [of sexual anomalies in humans] . . . always present themselves in animal conditions, that is, outside the reach of civilization, the sole purpose and effect of which is to restrain bestiality to the benefit of humanity" (Féré 1899, 87).

At this point, some readers may be wondering how the *Melolontha* story's farrago of outmoded attitudes and explanations could possibly be of interest to serious scientists or philosophers today. When it comes to (negative) attitudes toward animal homosexuality, things have indeed changed quite dramatically, at least in recent decades. In his famous compendium on animal homosexuality, biologist Bruce Bagemihl (1999, 88) charged that the twentieth-century history of research on animal homosexuality was a history of homonegativity, "a nearly unending stream of preconceived ideas, negative 'interpretations' or rationalizations, inadequate representations and omissions, and even overt distaste or revulsion toward homosexuality." Throughout the nineteenth and the twentieth centuries, observations of homosexual behavior in animals were indeed greeted with a variety of negative emotions and attitudes: disbelief, denial, denunciation, disgust, dismay, dismissal, fear, hatred, indifference, and outrage. In contemporary biology, however, this homonegativity has given way to a more neutral curiosity in the peculiarities of animal homosexual behaviors.[7]

Nevertheless, when it comes to explanations of animal homosexuality, the *Melolontha* story can still be useful today, for example, in casting doubt on the contemporary orthodoxy that nonhuman animals are only capable of homosexual behaviors, and that these behaviors should be explained in mechanistic rather than mentalistic terms. These issues are central in the next sections.

EXPLAINING AWAY ANIMAL HOMOSEXUALITY

Some contemporary zoologists would likely agree with Féré's mechanistic hypothesis that animal homosexual behaviors are nothing but "reflex phenomena or automatisms" (Féré 1899, 87), and that they are not to be associated with other dimensions of homosexuality, such as desires, preferences, or orientations (let alone identities). According to this view, animals only engage in homosexual behaviors if there are no opposite-sex partners available or if they fail to identify the sex of their partner—two conditions that, in turn, are often precipitated by other circumstances, such as uncontrollable lust (in the active partner) or exhaustion (in the passive partner). These two conditions correspond to two popular explanatory hypotheses, which we describe here as the *Hobson's choice hypothesis* and the *error hypothesis*, respectively. In the following, we scrutinize the historical and current importance of these hypotheses. At the same time, we assess Bagemihl's claim that we ought to dismiss them as "negative 'interpretations' or rationalizations" that try to explain away animal homosexuality (Bagemihl 1999, 88).

A Hobson's Choice

A first hypothesis for homosexuality in animals holds that animals only court or mate with same-sex partners if they do not have a choice, that is, in the absence of available and suitable partners of the opposite sex. Thus, homosexuality would be the outcome of a Hobson's choice—a free choice in which only one option is offered. The expression is named after Thomas Hobson, a sixteenth-century English stable owner who rented out his horses. To prevent his best animals from being chosen over and over again, he offered customers the choice of taking either the horse in the stall nearest the door or none at all. In the *Melolontha* story, Charles Féré was a prominent supporter of the Hobson's choice hypothesis. Concluding a summary of the experiments discussed above, he noted: "Homosexual relationships among cockchafers do not occur except in abnormal conditions. *The pursuit of another male only arises in the absence of females*" (Féré 1899, 84–85; italics ours).

Féré certainly was not the first to support this view. The Hobson's choice hypothesis was already quite popular in classical antiquity. In fact, classical literature abounds with observations of homosexual behavior in domesticated animals, particularly in poultry and birds. In his *Historia animalium*, Aristotle reported such behavior in partridges, quails, and cocks. Claudius Aelianus (1666, 1.15) spoke of "love-making" pigeon hens in a passage in *Varia historia*—a passage that translators have often coyly omitted. In his *Naturalis historia*, Pliny the Elder (1855, 10.79.58) reported that female pigeons, "if there should happen to be no male among them, will even tread each other, and lay barren eggs, from which nothing is produced." He also helpfully mentioned that the resulting "wind-eggs" were not particularly tasty. During the first centuries of the Common Era, animal homosexuality appeared to be a hot topic. "By the time of the early Christian fathers," the American historian John Boswell (1980, 152) claimed, "almost all zoologists considered some animals homosexual."

However, we must add nuance to Boswell's claim that animal homosexuality was common knowledge in antiquity. Most classical authors held the view that one can ascribe homosexual behaviors to animals but not homosexual desires or preferences.[8] The Roman poet Ovid is a good example. In his *Metamorphoses*, a lesbian protagonist complains about the absence of female-female desires, emotions, and (even) behaviors in animals:

> What way out is there left, for me, possessed by the pain of a strange and monstrous love, that no one ever knew before? If the gods wanted to spare me they should have spared me, but if they wanted to destroy me, they might at least have visited on me a natural, and normal, misfortune. Mares do not burn with love for mares, or heifers for heifers: the ram inflames the ewe: its hind follows the stag. So, birds mate, and among all animals, not one female is attacked by lust for a female. (Ovid 2000, book IX, 731ff)

If rams or mares mate with individuals of their sex, Ovid suggests, it is only because there are no ewes or studs available. The attentive reader will remember that Gadeau de Kerville would later conceptualize such behavior as "pederasty by necessity." Most classical authors considered Gadeau de Kerville's second type of pederasty, "pederasty by preference," a human exclusivity.

Plutarch's *Gryllus* illustrates this claim in the amusing debate between Odysseus and his former friend Gryllus who, on the island of Aeaea, had been transformed into a pig. Odysseus wants to turn his old friend back into his human form, unwilling to allow him "to grow old in the unnatural guise of beasts" (Plutarch 1900, book IX, 985E). Surprisingly, Gryllus fiercely resists, regaling his human visitor with an ode to animality—its moral virtues, reason, courage, and integrity. When it comes to the virtue of temperance, Gryllus argues, animals are far superior to humans: they have no eyes for gold and silver, they only eat when hungry, and they mate once a year. Males and females

> celebrate at the proper time a love without deceit or hire, a love which in the season of spring awakens, like the burgeoning of plants and trees, the desire of animals, and then immediately extinguishes it. Neither does the female continue to receive the male after she has conceived, nor does the male attempt her. So slight and feeble is the regard we have for pleasure: our whole concern is with Nature. Whence it comes about that to this very day the desires of beasts have encompassed no homosexual mating. But you have a fair amount of such trafficking among your high and mighty nobility, to say nothing of the baser sort. (990C–D)

There is no such thing as animal homosexuality, Gryllus claims, and if there is, it has nothing to do with lust and sexual preference but rather with "necessity" (991A). He provides the example of "a cock that mounts another for lack of a female" (990E). (And that cock was sent straight to the stake, Gryllus mutters, even though it had no choice.)[9] If animals are given a real choice, rather than a Hobson's choice, they will always prefer to have sex with opposite-sex partners. Human lewdness compares badly to this paragon of temperance: "Men do such deeds as wantonly outrage Nature, upset her order, and confuse her distinctions" (990F).

Throughout history, one finds a similar logic behind the many statements denying the existence of animal homosexuality. The well-known nineteenth-century forensic psychiatrist Johann Ludwig Casper, for example, assumed homosexuality to be a doubtful privilege of the human species: "To my knowledge there is no such thing anywhere in the animal kingdom, neither in males nor in females" (Casper 1863, 34). In our view, Casper meant "such thing" to refer to a homosexual preference, orientation, or identity.

The historical popularity of the Hobson's choice hypothesis is probably due in part to the fact that many of the relevant observations were based on sex-segregated domestic animals, particularly in zoos, private homes, and farms. When, at the end of the nineteenth century, review studies of animal homosexuality began to appear in scientific journals, field observations were practically absent. A typical example is Albert Moll's chapter on the topic in *Untersuchungen über die Libido sexualis* (1898). By his own account, he trusted the then director of the Frankfurt Zoo, entomologist Adalbert Seitz, even though Seitz's observations were strictly limited to captive species. One of these described an old billy goat that mounted some younger males despite the presence of several nanny goats. Seitz's own explanation is an instructive example of the Hobson's choice hypothesis, as he thought it likely that the females were pregnant and therefore unavailable (Moll 1898, 384).

We do not deny that the Hobson's choice hypothesis can help us explain a large number of homosexual behaviors in animals. However, we agree with Bagemihl in objecting to the casualness with which this explanation has been, and continues to be, used in the scientific literature. In our view, it would be rash to promote it as a default hypothesis. Once more, the *Melolontha* story provides some interesting clues. Féré believed, for example, that male homosexual behavior only occurred in the absence of females, but many of the observations he quoted, such as Peragallo's field report, specifically mentioned the availability of females. Moreover, all the cases of homosexual mating in Féré's own research experiments took place in the presence of sexually responsive females, which is the most striking aspect of the *Melolontha* story, as Gadeau de Kerville aptly observed (Gadeau de Kerville 1896a, 7; 1896b, 86). Finally, Féré knew about similar observations of homosexual mating despite the presence of opposite-sex mates. Like many other commentators, he referenced the work of Alessandro Muccioli, a contemporary Italian authority on pigeons. In a paper fashionably entitled "Degeneration and Criminality in Pigeons," Muccioli mentioned repeated observations of homosexual couplings in male Belgian carrier pigeons, even in the presence of available females. In fact, Muccioli's evidence prompted the famous sexologist and physician Henry Havelock Ellis to claim that "it is probable that sexual inversion in the true sense [i.e., Gadeau de Kerville's 'pederasty by preference'] will be found commoner among animals than at present it appears to be" (H. Ellis and Symonds 2008, 98).

In brief: some nineteenth-century observations suggested that homosexuality also occurred in noncaptive and sex-integrated animal populations, including phylogenetically remote species such as cockchafers. In later sections of this chapter, we indicate that recent zoological research corroborates this interpretation through both wildlife observations and experimental evidence of homosexuality in a variety of animal species.

A Woeful Mistake

A second hypothesis for homosexuality in animals holds that it involves some kind of misidentification or misunderstanding, particularly on the part of the penetrating partner. In the *Melolontha* story, Féré defended the error hypothesis, readily dismissing homosexual behavior among male cockchafers as an inevitable (and woeful) mistake. In his view, penetrating *Melolontha* males were lured into copulation by the presence of female pheromones on the body of the penetrated partner (who had just had sex with a female). The likelihood of such a mistake would increase under certain circumstances, such as hypersexuality of the active partner or fatigue of the passive partner.

We do not deny that the error hypothesis can help us explain some homosexual activities in the animal world. Bedbugs (*Cimex lectularius*) may be a good example. A recent study shows that male bedbugs have evolved alarm pheromones to signal their sex to same-sex partners, thus reducing the risk of homosexual mounting (Ryne 2009). Same-sex mounting in bedbugs appears to be highly traumatic for penetrated partners, as their abdomen is pierced by the needlelike penis of the penetrating partner. Females have evolved a secondary genital opening to reduce the cost of being wounded during insemination.

Penguin behavior similarly illustrates the potential value of the error hypothesis, as researchers often assume penguins score badly in sex recognition. In a recently discovered pamphlet about the sexual habits of Adélie penguins (*Pygoscelis adeliae*), originally written in 1911–12, the British Antarctic explorer George Levick noted that male penguins were somewhat casual about sexual morality. He reported many cases of adultery and sexual "hooliganism," where nonmating "hooligan cocks" sexually abused adult females and youngsters (Russell, Sladen, and Ainley 2012, 392). Some males even had sex with the bodies of dead females—an observation that was confirmed in the 1970s by David Ainley's experiments in another colony of the same species. Ainley (1978) froze a dead female's body in the typically submissive mating position and positioned it in a nest. He found that many breeding males attempted to mount the dead body, while the young males mostly limited themselves to a brief display, all of them apparently unaware of the condition of their sex partner. Levick concludes his pamphlet with an observation of mutual treading among male penguins: "Here on one occasion I saw what I took to be a cock copulating with a hen. When he had finished, however, and got off, the apparent hen turned out to be a cock, and the act was again performed with their positions reversed, the original 'hen' climbing on to the back of the original cock, whereupon the nature of their proceeding was disclosed" (Russell, Sladen, and Ainley 2012, 393). In one of the notebook entries upon which the pamphlet was based, Levick sighed: "There seems to be no crime too low for these Penguins" (Russell, Sladen, and Ainley 2012, 389).

A later Antarctic expedition in the 1930s reaffirmed Levick's observations and supplied an explanation for them. Polar expert Brian Roberts (1940, 213) hypothesized that penguins were "unaware of sex differences and do not differentiate between males and females even in mating." Many penguin species are sexually monomorphic, and males and females look very similar, if not identical, at least to the human eye. For decades, most scientists simply assumed that penguins themselves were also unable to recognize sex differences, and that mating misidentifications were the main cause of homosexual behaviors in this branch of the animal kingdom.

This instantiation of the error hypothesis was criticized, however, in a recent study on male-male courtship behaviors or "mating displays" in a large colony of king penguins (*Aptenodytes patagonicus*) on Kerguelen Island (Pincemy, Dobson, and Jouventin 2010). Over 25 percent of displaying pairs in the colony consisted of two males who courted each other on the lek during the breeding season with behaviors that also occurred in heterosexual displays. Standing close to each other, they stretched their bodies to a maximum height, rotated their heads in unison, and exposed various body parts. Why would they do that? The researchers ruled out (maybe too quickly) the error hypothesis, as their findings indicated that homosexual pairings occurred less often (and heterosexual more often) than one would expect on the assumption that males (and females) were pairing randomly and unable to detect their partners' sex (Pincemy, Dobson, and Jouventin 2010, 1214).

To explain these birds' behavior, the researchers finally gravitated toward a variant of the Hobson's choice hypothesis, indicating that even though their sample of displaying birds consisted of both males and females, it still had a sex ratio of 62 percent males. Therefore, some males on the lek did not have access to females. For those who might wonder why these males didn't simply walk away, hoping they would be luckier next time, the researchers provided an extra explanation, referring to earlier research that indicated that "when king penguins came back from the sea, males had high concentrations of testosterone and luteinizing hormone, *rendering individuals extremely motivated to display, perhaps to any adult king penguin*" (Pincemy, Dobson, and Jouventin 2010, 1214; italics ours). Once on the lek, and with these hormones already present, males cannot help but start courting whomever they meet. This mechanistic explanation is a good example of Féré's claim that animal homosexual behaviors are nothing but "reflex phenomena or automatisms" (Féré 1899, 87).

We are not really convinced by all aspects of this account. After all, a more cautious and empirically sounder conclusion would be that, even though this particular penguin species generally seems to be able to correctly

Figure 2.2: "Cold embrace." Two mounting male frogs in a mountain lake (2011). © Cyril Ruoso.

identify their partners' sex, mistakes might still occur. However, we agree with the authors of this study that it casts some doubt on the ubiquity and casualness with which the error hypothesis continues to be used to explain instances of animal homosexuality, both within and outside the realm of science (Bagemihl 1999; Roughgarden 2009). One example of this casualness can be found in a recent wildlife photography exhibition at the Natural History Museum in London. A prized photograph (fig. 2.2) depicts two male frogs holding on to each other in a mountain lake in the French Alps.

The photographer's commentary on the image recalls both the Hobson's choice hypothesis and the error hypothesis. He assumed, for example, that the two males were mating in the absence of a suitable female—a good example of the Hobson's choice hypothesis: "Here, one male is holding onto another as they both wait for a female to show up." He also appealed to the error hypothesis: "The clasping one may have mistaken the other for a female due to its larger size. I watched the error from my hide on the bank" (Ruoso 2011). The photographer's explanations may be correct, but they may also be yet another instance of people "explaining away something surprising by assuming the animals are somehow incapacitated" (Roughgarden 2009, 82).

Some contemporary commentators claim that both the Hobson's choice hypothesis and the error hypothesis are *negative* interpretations or even rationalizations of animal homosexuality, in that they tend to explain away the phenomenon. In philosophy, explaining away a phenomenon X usually means showing that X does not exist, or at least that it is no longer rational to believe that X exists (Ratzsch and Koperski 2005). One can do so by appealing to the explanatory power of alternative hypotheses in which X does not play any role, similar to how the caloric theory of heat was explained away by the mechanical theory of heat. In this context, "X" stands for animal homosexuality, and therefore the claim that some hypotheses explain away animal homosexuality can be interpreted as indicating that such a phenomenon does not exist. This conclusion, however, would be a bridge too far for most biologists. Rather, by providing mechanistic explanations for animal homosexual behaviors, that is, explanations in terms of sex ratios and hormone levels, they avoid mentalistic explanations, that is, explanations in terms of certain mental states that underlie homosexual behaviors. In humans, such behaviors are usually seen as indicative of an underlying homosexual desire, preference, or orientation. So what is being explained away by mechanistic hypotheses, such as the Hobson's choice hypothesis and the error hypothesis, is the existence of animal homosexual behavior qua expression of underlying mental states, in addition to the existence of homosexual mental states as such (which may or may not be expressed in behavior). If anything, the *Melolontha* story and many contemporary case studies on animal homosexuality illustrate how difficult it is for animal scientists (and laypeople) to even entertain the possibility that animal behavior is guided by mental states like sexual desires or preferences.

Indeed, the essential message of negative hypotheses like the error hypothesis or the Hobson's choice hypothesis seems to be that animal homosexuality is not "real" homosexuality, since it does not relate to, or result from, particular homosexual mental states. Animals do not *desire* to mate with same-sex partners; they are simply being deluded or forced by circumstances. Surely, they do not *prefer* homosexual activities either. Given a real choice, they will prefer a partner of the opposite sex. It should not surprise us, then, that biologists have often preferred to describe homosexual interactions as "pseudo-copulations," "mock courtships," "sham matings," "false mountings," "pseudo-pairs," and so on (Bagemihl 1999, 96–97; see also Despret 2016). Later in this chapter, we will see that, for various reasons, some biologists even go so far as to doubt the propriety of the term "animal homosexuality" in an attempt to safeguard it for humans.

But first things first: What does it mean, for humans and other animals, to desire or prefer homosexual behaviors and activities? What does it mean to have a homosexual orientation or identity? In the next section, we examine the conceptualizations of these mental states in turn, providing definitions and operationalizations where possible. Our main argument is that it does make sense to ascribe homosexual desires, preferences, and orientations to at least some animal species.

HOW GAY CAN YOU BE?

Critics of animal homosexuality often charge those who use the term with anthropomorphism. Anthropomorphism implies a (facile) projection of human characteristics onto nonhuman animals. This projection is based on the assumption that both species are relevantly similar, at least in certain respects, and it often serves to confirm that similarity.

Critics have used the anthropomorphism objection in at least two different ways (following Fisher 1996). *Categorical* anthropomorphism argues that it would be a category mistake to ascribe homosexual mental states to nonhuman animals because these animals are not capable of having the mental states that are usually associated with human homosexuality, such as desires, preferences, orientations, and identities. This particular objection leads us to consider the nature of these states, as well as the question why nonhuman animals would not be able to have them. A second kind of anthropomorphism, *situational* anthropomorphism, argues that at least some animal homosexual behaviors and mental states should not be labeled as homosexual because they are not sexual in the first place, even though the animals may be capable of having homosexual mental states.

Categorical anthropomorphism implies that ascribing human characteristics to nonhuman animals amounts to a category mistake. A category mistake is a fallacy in which something of a specific type or logical category is considered to belong to a different type or logical category. Following Gilbert Ryle's famous example, a tourist commits such a fallacy when, after a guided tour of the many university buildings in Oxford, he asks to be shown the university itself (Ryle 1949). The mistake arises from the assumption that a university belongs to the logical category of buildings, rather than to the logical category of organizations or institutions. Similarly, it would be misguided to ascribe homosexual desires, preferences, orientations, or identities to cockchafers or king penguins because they do not belong to the logical category of beings capable of having such mental states.

The likelihood that a nonhuman animal species has a mind like ours is generally inversely proportional to its phylogenetic distance from us. Therefore, it is rather implausible to attribute highly complex mental states to nonhuman animals that, from a phylogenetic point of view, are far removed from the human species, like insects. It does not seem plausible, for example, to say that common cockchafers know that we know that they belong to the family Scarabaeidae. In philosophical parlance, such highly complex mental states are known as second-order beliefs. But what about desires, preferences, orientations, and identities? To answer this question, we need to specify what constitutes a desire, a preference, an orientation, and an identity and how these concepts are to be operationalized in animal research.

Desire

One way of ascertaining whether an organism exhibits a homosexual desire is to check for bodily signs of sexual arousal in the context of a homosexual encounter. In most male mammals, one can determine sexual arousal by visual inspection of the genitalia or by means of a technique known as penile plethysmography, which measures penile blood flow. These methods have limited applications, however, since not all male animals have penises. For instance, the males of many bird species lack a penis. Another important limitation of this approach is that bodily arousal represents only one aspect of sexual desire that, in humans, does not always correlate with a more important aspect of desire, subjective arousal. In humans, nocturnal penile tumescence (or a morning erection) is a common phenomenon that does not (always) involve subjective arousal (Singer 1984). Consequently, subjective arousal in humans is usually measured through questionnaires that ask participants whether and to what extent they are "turned on," "find pleasure," or "feel sexually aroused" by relevant stimuli (Rellini et al. 2005).[10] Such questionnaires cannot be used in the study of animal sexuality, so how are we to find out whether animals have sexual desires?

Philosophers have long debated the many ways to define the concept of desire. Some philosophers have focused on subjective arousal and feelings of pleasure that tend to accompany the satisfaction of a desire (G. Strawson 1994). Other definitions emphasize the distinctive motivational value of desires, with some philosophers arguing that having a desire essentially means one is disposed to act in ways that are likely to bring about something (the object of desire, a desired state, and so on) (M. Smith 1994). According to this view, desires are dispositions to act, and animal scientists have attempted to operationalize such dispositions by measuring—often quite literally—the lengths to which an animal will go to obtain sexual contact with a physically distant potential mate that is only available via visual, auditory, or olfactory stimuli (Pfaus, Kippin, and Coria-Avila 2003). Many studies have shown, for example, that male rats are willing to sustain multiple electric shocks to have sex with another male that they perceive to be in a typical mating posture (Pfaff 1999; Pfaus et al. 2012). Similar behavioral patterns have also been found in males of many other vertebrate species, including guinea pigs, pigeons, and sticklebacks (Pfaus et al. 2010). If these studies are sound, it seems fair to conclude that the animals involved possess motivational structures that are somewhat similar to human sexual desires, and it is therefore not necessarily a category mistake to ascribe sexual desire to nonhuman animals.

Admittedly, operationalizing sexual desire as a disposition to act would gain a fair number of false negatives in the context of research on human male sexuality. A false negative occurs when a condition, such as a desire, is considered absent when it is in fact there. A false positive represents the reverse problem, where a condition is considered present when it is not there. Typically, dispositions do not always turn into actual behaviors. Some men are too shy, too calculating, or too conscientious to pursue their every sexual desire. In most nonhuman animals, excluding primates, such factors are probably relatively unimportant, so the chances of false negative results are rather slim. This is an important point, especially about what typifies *human* sexual desire. The number of false positives will be limited. Even in our species, most individuals who choose to have sex are indeed driven by a sexual desire, and there is no clear reason why false positives would be more abundant in other species. A cautious conclusion would be that the motivational account (and its operationalization of desire as a disposition to act) is a reliable instrument, though not a litmus test, to ascertain the presence of homosexual desires in animals.

Still, some people will disagree with this conclusion. They might say that the above analysis and operationalization of sexual desire is too liberal because it allows animal scientists to attribute complex mental states, including desires, to animals like insects. Can cockchafers have sexual or homosexual desires? Some philosophers argue that sexual desires of such animals, if they have any, cannot be all that similar to human sexual desires because the animals in question lack the cognitive machinery that humans require to produce such complex mental states. Edward Stein, for example, employs this argument to criticize the use of fruit flies as models for human homosexuality: "Fruit flies cannot be said to have desires or to experience attraction in ways even remotely like how humans have desires and experience attraction. . . . [H]uman sexual responses are cognitively mediated, by which I mean that human sexual desire is intimately intertwined with our thinking processes. Flies do not have the relevant thought processes; in fact, they are at best borderline cognitive systems" (Stein 1999, 166–67).

To be clear, Stein argues not just that fruit flies do not have desires like ours but also that fruit flies do not have desires at all. Moreover, in Stein's view, it is very likely that birds do not have sexual desires either: "Although birds certainly have greater cognitive capacities than flies, it is not clear whether seagulls have the cognitive architecture required for desires and intentions; after all, the expression 'bird brain' means 'stupid' for a reason" (Stein 1999, 170–71).[11]

Stein is not alone in arguing that, conceptually speaking, desires cannot be detached from beliefs or, in his own words, "thinking processes." Indeed, some philosophers think that a desire should be equated with the belief that the object of the desire is good. This would be a third view of the nature of desire, alongside views that define desire in terms of subjective arousal or behavioral dispositions. If such good-based theories hold weight, then animals cannot have desires unless they can have beliefs, another topic that philosophers tend to disagree about (D. Davidson 1982; Rowlands 2009).

In our view, Stein's analysis of the concept of desire is overly restrictive. Of course, human sexual desires are cognitively mediated, but so are many other human behavioral patterns. Human attitudes and emotions are intimately connected with various thinking processes, but that does not prevent us from ascribing hunger and anxiety, for example, to at least some animal species. In doing so, we do not have to assume that human hunger is identical to animal hunger, only that both phenomena are similar enough. Likewise, speaking of homosexual desires in cockchafers and king penguins does not automatically equate them with human homosexual desires. Admittedly, unlike animal sexuality, human sexual desire may often involve "a desire that one's partner be aroused by the recognition of one's desire that he or she be aroused" (Nagel 1969, 12), but we see no good reason to turn this interesting aspect of human sexuality into a prerequisite for the use of "sexual desire" as a label.

Stein's analysis, however, is not just overly restrictive. It is also presumptuous in suggesting that there is a widespread consensus among both scientists and philosophers about what constitutes a desire. In reality, the concept of desire is just one of a myriad of concepts, including the concepts of knowledge and justice, that philosophers and scientists agree to disagree about (Silverman 2000). Like many other concepts, the concept of desire originates in a body of knowledge philosophers and psychologists refer to as folk psychology—a mishmash of lay constructs and conceptions, all of which have been endlessly adapted and rearranged throughout history in order to meet the changing demands of their users. Therefore, it should not surprise us that "desire" has evolved to refer to a whole range of phenomena, and that no single definition can capture the necessary and sufficient conditions for the correct application of this concept. Phenomena as divergent as thirst, homosexuality, and philosophy can all legitimately be considered as instances of desire—craving whiskey, wanting sex, and longing for wisdom. Given the origin of these concepts, it is unlikely that philosophers and scientists will ever be able to define them in a way that finds favor among even a small majority of their colleagues. Stein seems to ignore this dissensus when chastising animal scientists for ascribing sexual desires and preferences to nonhuman animals. The point is, however, that there are many definitions of desire and preference, that some of these definitions allow us to discuss homosexual desires and preferences in some animal species, and that there is no benchmark for deciding which of these definitions is most reliable or valid.

At this point, anti-anthropomorphists might concede that some definitions of desire are useful when discussing homosexual desires in nonhuman animal species. At the same time, they would insist on reserving the term "homosexuality" for human behaviors and mental states, because at least some of these behaviors and mental states cannot be found elsewhere in the animal kingdom. Their argument is then that while male cockchafers and king penguins may be capable of sexually desiring one another, their behaviors should not be described as homosexual, for some other mental state is the only true kind of homosexuality, which, unlike behaviors and desires, can only be found in the human species.

Preference

One example of this supposedly uniquely human mental state would be a homosexual preference. For better or worse, the history of research on animal homosexuality reminds us that a sexual desire per se does not provide evidence of a sexual preference. Males engaging in homosexual behavior may have mistaken their sex partner for a female (the error hypothesis), or they may not have had a choice between a male and a female in the first place (the Hobson's choice hypothesis). In Gadeau de Kerville's vocabulary, these behaviors would be considered instances of "pederasty by necessity." And even though such sexual behaviors may well be motivated by a homosexual desire,[12] they do not necessarily attest to a homosexual preference. Only males who are shown to mate with males in the presence of available females would be considered "pederasts by preference." Gadeau de Kerville's distinction also seems to be true of human male homosexuality. Many studies have reported, for example, that homosexual behaviors among men are rampant in the army and in certain prison populations, even though many of the men who freely engage in homosexual behaviors and report having homosexual desires still claim to prefer heterosexual sex to homosexual sex (Hensley and Tewksbury 2002; Hekma 1991).

In the past few decades, the work of primatologist Paul Vasey has substantiated and operationalized Gadeau de Kerville's analysis. Vasey proposes we should speak of homosexual preferences only in contexts where a male prefers to have sex with another male rather than an available female. A preference always entails some kind of choice, and the choice involved needs to be more than just a Hobson's choice. Both options have to be equivalent to the extent that both may result in sexual contact. Vasey provides a list of five criteria to determine whether a given animal exhibits a homosexual preference rather than simply a homosexual desire: "First, the subject should be able to *simultaneously choose* between a male or female. Second, the two stimulus animals should ideally be *sexually proceptive* vis-à-vis the subject. Third, these interactions must *culminate in actual sexual behavior* between the subject and the same-sex stimulus animal. Fourth, the subject must be *uncoerced*. Fifth, it must be demonstrated that *the behaviors used to measure sexual partner preference are sexual, at least in part*" (Vasey 2002, 147; italics in original).

Research on male-male tandem formation in the blue-tailed damselfly (*Ischnura elegans*) suggests there is indeed homosexual preference, in Vasey's sense, in the nonhuman animal kingdom, even among "lower" animals like insects. (Rams are another, perhaps more classic, example, as 9 percent of these animals prefer same-sex partners [Roselli 2020].) Tandem formation is a typical courtship behavior in various damselfly species. It consists of one individual clasping another at the prothorax—the front segment of the trunk—with its abdominal appendages. The clasped individual decides whether copulation ensues after the tandem is formed—a decision most likely based on an assessment of the relevant characteristics of the grasping individual.

Similar to same-sex courtship in king penguins and same-sex mounting in cockchafers, male-male tandem formation in damselflies appears to be rather common in wildlife populations where both males and females are abundantly available (Utzeri and Belfiore 1990). More recent experimental evidence has confirmed these earlier wildlife observations (Van Gossum, De Bruyn, and Stoks 2005). Nearly 20 percent of freshly caught male blue-tailed damselflies preferred to form a tandem with another male in a binary-choice experiment wherein each (focal) male individual was given the choice between a male and a female. To their surprise, scientists also found that nearly 75 percent of males displayed what they saw as a homosexual preference after they had spent a few days in an all-male insectary. The numbers declined again after the same males were immersed in an all-female population.

Critics might object that these experiments do not prove the existence of a homosexual preference in some damselflies. The most obvious alternative explanation for this behavior is that they are unable to recognize sex differences—an inability that would easily lead to mating misidentifications. In this case, however, the error hypothesis does not hold. To be sure, some female damselflies resemble male congeners in body coloration. In the literature they are called andromorph females, as opposed to gynomorph females, which have a distinctively female coloration (Bots et al. 2009; Van Gossum et al. 2008). To rule out sex-recognition issues in their research on homosexuality in damselflies, scientists organized two parallel binary-choice experiments: one in which males had to choose between a male and an andromorph female, and another in which they had to choose between a male and a gynomorph female. The results were highly similar: in both experiments, approximately 20 percent of males displayed a homosexual preference (Van Gossum, De Bruyn, and Stoks 2005).

The damselfly story also resists the Hobson's choice hypothesis, which states that homosexual behavior in animals occurs if and only if suitable partners of the opposite sex are unavailable. In both wildlife and experimental conditions, a significant percentage of male damselflies preferentially court other males, even when females are available—thus lending evidence to Gadeau de Kerville's 1896 hypothesis that there is such a thing as homosexuality by preference in the insect world.

Finally, this example reveals a limitation in Vasey's analysis and operationalization of the concept of animal homosexual preference. In our view, Vasey's list of criteria lacks an element of continuity or stability. A strict interpretation would allow us to conclude that a single sexual choice reveals a sexual preference. This conclusion seems counterintuitive. Our instinct says that ascertaining a sexual preference involves more than a simple snapshot; it also involves a given choice to be repeated over time in consecutive sexual interactions, preferably in different social and physical environments, where both males and females are simultaneously available. This critique also applies to the aforementioned damselfly experiments, in which the researchers failed to monitor the stability of sexual choices throughout consecutive experiments. It would indeed be interesting to know whether those males who preferentially courted other males also displayed a homosexual preference in later experiments, after they were exposed to all-male or all-female populations.

Orientation

A solution to the problem with Vasey's definition of preference would be to distinguish between a homosexual *preference* and a homosexual *orientation*, reserving the term "preference" for transient or even single sexual choices and the term "orientation" for more continuous or even permanent homosexual preferences that, by definition, tend to exclude heterosexual

Figure 2.3: Homosexual copulation in bottlenose dolphins (1999). © John Megahan.

activities. The twin elements of stability and exclusivity make it very difficult for wildlife biologists to ascertain a permanent homosexual orientation in nonhuman animals. After all, it is virtually impossible to determine whether an animal in the wild has never had any heterosexual contact, desire, or preference, let alone whether it will never have one in the future. It shouldn't surprise us, then, that some scientists consider homosexual orientations to be "very uncommon" in nonhuman animals (see, e.g., LeVay 1996, 270).

If, however, we exchange this thick concept of homosexual orientation for a slightly thinner version, defined, for example, as a relatively stable sexual preference in a same-sex companionship or series of companionships,[13] then homosexual orientation does in fact exist, and in various classes across the animal kingdom. Male bottlenose dolphins (*Tursiops truncatus*) are perhaps the best-known mammalian example. On this species, Bagemihl (1999, 48) notes: "The majority of males in some populations form lifelong homosexual pairs, specific examples of which have been verified as lasting for more than ten years and continuing until death" (see fig. 2.3).[14]

In other animal classes, stable, long-term companionships that involve sexual behaviors such as mounting or courtship have been documented more extensively. The literature abounds with observations of such companionships in bird species, including roseate terns (*Sterna dougallii*) (Nisbet and Hatch 1999) and Laysan albatrosses (*Phoebastria immutabilis*) (Young, VanderWerf, and Zaun 2008). Importantly, none of these cases are associated with sex segregation. They all involve animals who have access, either continually or occasionally, to members of the opposite sex.

Identity

One last stronghold of anti-anthropomorphism in the context of animal homosexuality is the concept of homosexual identity. While this concept plays an important part in numerous psychological, historical, and sexological studies about human homosexuality, it has always been alarmingly underdefined, as the psychologist and philosopher Vivienne Cass (1984a, 107) noted in the 1980s: "There are literally hundreds of scientific articles that refer to homosexual identity, without explaining what is meant by the concept." It is quite telling, for example, that a prestigious handbook like *The Lesbian and Gay Studies Reader* (Abelove, Barale, and Halperin 1993) contains plenty of essays about sexual identities and identity politics without ever actually explaining what the concept of identity is supposed to mean.

This neglect is regrettable because the concept in question seems to be more ambiguous than most scholars assume. In some contexts, the term refers to a person who identifies in a social context as having either predominantly homosexual desires and preferences or a homosexual orientation. People signal their homosexual identity to others, then, by adopting a label that relates to their sexuality; this can be one label among a host of many others. Sometimes, however, the concept of homosexual identity can also play a rather political role, such as when someone identifies first and foremost as a homosexual person rather than as, say, a person with a particular ethnicity or gender. In other contexts, the concept of homosexual identity primarily denotes a set of properties that are somehow associated with homosexuality and that are considered characteristic of someone with a homosexual preference or a homosexual orientation in given social situations (or in a particular society). For example, some individuals adopt a particular lifestyle because they see themselves as homosexuals (Cass 1984b; Troiden 1985, 1988), and this lifestyle is seen as part of a gay or homosexual identity. In the psychological literature on homosexuality, however, the term "identity" is occasionally also used to refer to people who simply answer "gay" or "homosexual" to the question of what their sexual orientation or preference is.

One common denominator in these different definitions is that a homosexual identity seems to presuppose self-consciousness or self-awareness; it is one way of answering the question "Who am I?" (Howard 2000). This condition would appear to prevent most animal species from developing even a vaguely similar characteristic to a homosexual identity in humans. Some animal scientists claim to have found indications of both self-consciousness and homosexuality in certain primate species such as gorillas, orangutans, and chimpanzees, as well as in bottlenose dolphins, killer whales, elephants

(Plotnik, de Waal, and Reiss 2006), and even magpies (Prior, Schwarz, and Güntürkün 2008), all of which appear to recognize themselves in their reflection in a mirror.[15] Still, none of these scientists go so far as to infer that some animals identify with their sexual orientation (Vasey 2002; N. Bailey and Zuk 2009; Poiani 2010). It seems fair to say that the presence of both homosexual preference and self-consciousness is probably a necessary, but certainly not sufficient, condition for an organism to exhibit a homosexual identity, as is testified in the sexual history of many human cultural groups.

Occasionally, however, identity talk causes confusion in the debate about animal homosexuality. According to Jennifer Terry, for example, one particular study of homosexuality in fruit flies (Odenwald and Zhang 1995) was based on the assumption that "mutant flies" can be classified "in terms of identity—that is, as gay fruit flies rather than simply as flies that exhibited homosexual behaviour" (Terry 2000, 167).[16] Terry's interpretation was not very charitable, however, and was largely based on a misinterpretation of the authors' use of the term "gay." In the original study, Odenwald and Zhang did not use that word to suggest that fruit flies self-identify in terms of their sexual behaviors, desires, or preferences. They simply, though perhaps unwisely, considered it a synonym for "homosexual."

In sum, one can only continue to use the categorical anthropomorphism objection against animal homosexuality by equating "homosexuality" with "homosexual identity." While there are no principled objections in doing so, it does go against the practice, both within and outside the animal sciences, of labeling at least some animal activities, desires, preferences, and orientations as homosexual.

SEX OR DOMINANCE?

Even if one accepts that many animals are minded creatures, and that they are able to have some similar mental states to those we associate with human homosexual behaviors, one could still argue that attributing homosexuality to nonhuman animals is not so much a categorical as a factual mistake. According to this argument, when talking about animal homosexuality, homosexual behaviors and mental states are perceived where there are in fact none, because, for example, these acts are not sexual in the first place.

This argument appeals to a second kind of anthropomorphism, situational anthropomorphism, in which we erroneously ascribe a behavior or mental state to a nonhuman animal that may well be capable of such states but does not, at this moment, find itself in any of them (Fisher 1996). Typical examples of situational anthropomorphism include the grinning of chimpanzees, which humans often read as an expression of joy but is actually an anxious reaction to an unexpected and frightening stimulus. Chimpanzees may well be able to experience joy; they simply do not express it by means of grinning. Similarly, one could argue that male cockchafers and king penguins may, in principle, be capable of having homosexual desires and preferences, but that the observed mounting and courting in male cockchafers and king penguins should not be considered as manifestations of such mental states. These behaviors are not sexual, or so the argument goes, because they do not express a sexual desire or preference. Rather, they are manifestations of some nonsexual desire, such as the desire to collaborate with or to dominate other males.

Indeed, there are biologists who believe that some animal behaviors that are mistaken for homosexual behaviors function as social behaviors. More specifically, they are part of a power struggle within the social hierarchy of a given species. These behaviors help negotiate power relations, particularly among male individuals. It is not that difficult to read some version of this hypothesis into the *Melolontha* story. August Kelch, the very first naturalist to observe homosexual coupling in cockchafers, conjectured that "the larger and stronger of the two had forced itself on the smaller and weaker one, had exhausted it and only because of this dominance [*Ueberlegenheit*] had conquered it" (Kelch 1834, 737–38). Still, the question is why the dominant male wanted to engage in this behavior in the first place.[17]

In the second half of the twentieth century, the dominance hypothesis was elaborated on in order to answer this question. According to this theory, same-sex couplings are thought of as ritualized gestures to communicate and further establish one's rank within a dominance hierarchy, or a ranking system that determines one's access to resources and mates (Wickler 1967; Vasey 1995). The roles that individuals adopt during same-sex interactions reflect their position in the ranking system: penetrating expresses (and strengthens) one's dominance over the other, while being penetrated expresses one's subordination.

We agree that some cases of homosexual behavior in nonhuman animals can be successfully explained as an instrument to signal, express, establish, or strengthen dominance relationships. However, there are a few problems with this hypothesis. First, in some animal species, subordinate individuals mount their dominant partners more often than vice versa (Bartoš and Holečková 2006), much like when nineteenth-century entomologists sometimes found smaller beetles on top of larger and stronger specimens. Second, the dominance hypothesis can only account for same-sex interactions that lend themselves to expressing dominance or submission, such as mounting or pelvic thrusting. Such activities, however, are only a limited subset of the same-sex behavioral repertoire. Mutual masturbation or courtship behaviors, which are also observed in quite a few nonhuman animals, do not fit in this account.

More importantly, some critics claim that the dominance hypothesis has played a significant part in a project that they describe as "desexualising same-sex sexuality" (Bagemihl 1999, 687; see also Sommer, Schauer, and Kyriazis 2006, 265) and that they interpret as yet another attempt to explain away animal homosexuality. A particular animal behavior is not homosexual behavior, or so the hypothesis goes, because it is not sexual behavior in the first place.

Yet how can homosexual interactions not be sexual interactions? The dominance hypothesis is one of many evolutionary hypotheses about animal homosexuality (for an overview, see Adriaens and De Block 2016). One of the questions driving these hypotheses is how does homosexuality contribute to the probability of the individual's survival and reproductive success. In Bagemihl's view, nearly all evolutionary explanations of animal homosexual behavior are guilty of desexualizing such behavior. Their underlying disjunctive logic dictates that it is either functional behavior or sexual behavior (see also Sommer, Schauer, and Kyriazis 2006, 265). In some versions of the dominance hypothesis, homosexual behavior is thought of as dominance behavior rather than sexual behavior. According to Ford and Beach (1951, 138), for example, Solly Zuckerman's work on female-female mounting in baboons implied that such behavior "probably constitutes an expression of dominance *rather than sexual desire*" (italics ours).

A recent case study illustrates the continuing allure of such reasoning. In August 2017, the British wildlife photographer and conservationist Paul Goldstein caught two male lions on camera "sneaking off into the bushes... for some privacy," somewhere in a far corner of the Maasai Mara National Reserve in Kenya (Malm 2017). His pictures document how one lion mounts the other in a copulation that takes approximately a minute— much longer than the average heterosexual lion copulation (see fig. 2.4a–b). After that, the couple hung around for a while, cuddling and nuzzling each other, and even, as Goldstein stated in a *Daily Mail* interview, "throwing a conspiratorial wink" (Malm 2017).

Goldstein had no problem considering the lions' behavior as homosexual, but some scientific commentators, like Craig Packer from the University of Minnesota Lion Center, were quick to dismiss his interpretation as an example of situational anthropomorphism. In an interview with the website Live Science, Packer claimed that "the mounting behavior isn't actually sexual" (Pappas 2017). Rather, he argued, "it's a social interaction that has nothing to do with sexual pleasure." Packer seems to be explaining away this homosexual behavior as social behavior instead of (homo)sexual behavior.

Figures 2.4a and 2.4b: "Gay" lions in the Maasai Mara National Reserve (2017). © Paul Goldstein.

To confuse matters more, the term "socio-sexual" often pops up in this context, the true meaning of which, according to Bagemihl (1999, 116), is "not fully or exclusively sexual." He concludes: "In the vast majority of cases, these functions are ascribed to a behavior instead of, rather than along with, a sexual component. . . . [T]he erasure by zoologists of sexual interpretations from same-sex contexts has been categorical and nearly ubiquitous" (Bagemihl 1999, 115–16).

On this subject, however, we strongly disagree with Bagemihl. First, let us again return to the *Melolontha* story. Quite clearly, nineteenth-century naturalists did not try to desexualize their cockchafers' coupling, indulging, as they did, in lengthy discussions about the technique and locus of same-sex penetration and documenting numerous examples of homosexual ejaculation in various animal species (Karsch 1900). As Peragallo (1863, 663) succinctly noted, penetration (in male-male cockchafer copulation) was often "fully complete, certain, and even intentional."

Second, even if some earlier scientists have made dubious statements about the sexual nature of animal homosexuality, the same cannot be said for the entire (evolutionary) biological literature on the topic. Most zoologists and evolutionary biologists are well aware of the possibility that homosexual behavior can serve one or more evolutionary functions, such as dominance expression or alliance formation (see chapter 3), while still having a sexual character. Homosexuality, whether human or animal, can be both functional and sexual (Sommer and Vasey 2006, 13; see also Poiani 2010). One can make a similar point by referring to the widely used distinction between ultimate (or evolutionary) and proximate explanations of organismic properties and behaviors. Even though the ultimate cause of heterosexual sexual activities relates to reproduction, only a very small percentage of these activities are motivated by a desire to procreate.

A third and final problem with Bagemihl's wholesale critique is its disregard for the possibility that (some) scientists do not consider (some) animal behaviors as sexual simply because these behaviors fall outside their implicit or explicit definition of sexuality. Suppose we narrowly defined sexual behavior, much like the protagonists in the *Melolontha* story, as "behavior involving genital penetration and including ejaculation." In that case, we would have good reason to not classify kissing, genital nuzzling, or beak-genital propulsion as sexual behaviors. (Beak-genital propulsion is a behavior typical of some species of waterfowl in which one animal inserts its bill into the anal opening of the other, thus propelling it through the water.) Therefore, in order to evaluate Packer's claim that same-sex mounting in lions is "not really sexual," we first need to understand how he defines sexuality and how the larger community of animal scientists has come to understand the term.

GRAB THAT PENIS!

What is sexuality? What exactly do we mean when we qualify a behavior or mental state as sexual? The question is a philosophical one about the nature of sexuality. In this section, we examine some answers to this question and discuss how they bear on the debate about animal homosexuality.

The philosophical literature about the nature of sexuality roughly breaks up into two camps, both of which have their successes and shortcomings: the extrinsic view of sexuality and the intrinsic view of sexuality.

Some philosophers have argued that activities and mental states are sexual to the extent that they contribute to a particular end, such as reproduction, love, or simple communication. Activities and mental states are not sexual in and of themselves; they derive their sexual character from the end they serve. This can be termed the *extrinsic* view of sexuality. In the mid-1970s, for example, Robert Solomon claimed that sex is a means of communicating a wide variety of feelings, ranging from love and anger to trust (Solomon 1975). Further back in history, the extrinsic view has been immensely popular, particularly as it allowed moralists to condemn as unnatural or perverse all kinds of activities and mental states that did not fulfill some sort of "higher purpose," such as love or reproduction (Goldman 1977).

One of the main problems with the extrinsic view is that it often confuses debate about the nature of sexuality with debate about the moral value of sexuality. However (historically) successful it has been in matters of morality, the extrinsic view fails rather miserably as an account of the nature of sexuality, if only because it brands any activity without the required end as not sexual. If sex is supposedly only about reproduction, for example, then oral and anal sexual activities are in fact not sexual at all, which is a strongly counterintuitive conclusion. One may argue that these activities are unnatural or perverse, but that does not mean they are not sexual.

More recently, many philosophers have defended the opposite view, which holds that sex is in the nature of particular activities and mental states themselves, rather than in the purpose they serve. This can be termed the *intrinsic* view of sexuality. As in the case of the extrinsic view, there are some variants within this position. One narrow variant focuses exclusively on genitals, considering all genital contact as sexual, regardless of the context in which it occurs. Another related view focuses on bodily arousal, as manifested in phenomena like rising penises, wet vaginas, hard nipples, and flushed skin. Still another variant links sexuality with psychological arousal. One of its supporters is Edward Stein, who claims, "A desire is sexual insofar as it involves... the [psychological] arousal of the person who has the desire and... a behavior is sexual insofar as it involves... the arousal of at least one of the participants in a behavior" (Stein 1999, 69). Another adherent of the intrinsic view is Thomas Nagel (1969), who argued that sex is all about two people being aroused by each other, who are aware of the other's arousal and aroused, in turn, by this awareness. A fourth variant of the intrinsic view claims that activities and mental states are sexual to the extent that they provide a particular kind of pleasure. According to Robert Gray (1997, 59), for example, "those activities... are sexual which serve to relieve sexual feeling or, alternatively put, which give rise to sexual pleasure" (see also Goldman 1977).

Philosophers being philosophers, they have not been able to reach a consensus on which variant of the intrinsic view best captures our commonsense intuitions about what is sexual and what is not. (There seems to be some consensus, though, that the extrinsic view generally fares worse than the intrinsic view.) Some variants are more useful for life scientists, as they allow them to objectively identify sexual activities and states not only in humans but also in a variety of other animal species. One can measure physical arousal by means of a plethysmograph, for example—a device that measures blood flow in the penis or the walls of the vagina. However, the trouble with the arousal view, which we have already alluded to in this chapter, is that bodily arousal also occurs in contexts that we generally do not consider to be sexual. Nipples also become erect during breastfeeding or in the cold, for example, and morning erections are not typically considered a sexual phenomenon. The genital view has the advantage of simplicity, but it is also quickly problematized, given that we do not consider gynecological and andrological examinations as sexual activities, even though they involve plenty of genital manipulation. Conversely, kissing and various other activities can qualify as sexual in some contexts, even though they do not involve genital manipulation.

The main problem that concerns most of the intrinsic view's variants is that they largely shift the question away from the nature of sexuality to the nature of (sexual) arousal or (sexual) pleasure. Stein, for example, does not define his concept of psychological arousal beyond claiming that "there is some abstract property or set of properties that mark off states of arousal [i.e., the psychological states of being aroused] from other states" (Stein 1999, 70). Similarly, Gray does not define his concept of (sexual) pleasure, nor does he attempt to distinguish it from other kinds of pleasure.[18] Finally, both Stein's and Gray's variants make it more difficult to identify sexual behaviors and mental states in nonhuman animals, as their cognitive and affective states are less accessible to humans. (However, as we argued earlier in this chapter, some concepts of some homosexual mental states have in fact been operationalized for the animal sciences, with some success.)

To better grasp the issue, let us consider a case study. Diddling is an interesting example of an animal behavior of which the sexual nature has been disputed. Primatologist Timothy W. Ransom (1981) coined the word "diddling" to refer to an asymmetrical kind of genital touching among adult male olive baboons (*Papio anubis*). More specifically, it involves one male grabbing or rubbing the other male's penis or fondling his scrotum. The latter male does not return the gesture, at least not immediately—hence the asymmetric nature of the behavior. A series of gestures and friendly facial expressions precedes the actual technique, including a swinging gait, smacking of the lips, and narrowing of the eyes. The whole sequence rarely lasts longer than a few seconds, during which both partners stand stock-still—a remarkable pause in the somewhat hyperactive behavior pattern of these animals (Smuts and Watanabe 1990, 152). Diddling has not been reported in animal species other than baboons (Whitham and Maestripieri 2003, 848).[19]

Diddling is part of a portfolio of ritualized greeting behaviors in several baboon species—a portfolio that includes mounting, embracing, genital nuzzling, and posterior presenting and grabbing. Baboon greetings constitute the sole setting in which adult males suspend their hostilities and feel out each other's willingness to engage in future coalitional or cooperative behavior. Indeed, the evolutionary function of these behaviors seems to be "to establish a kind of temporary 'truce' or 'neutral ground' on which relations between the greeters can be explored, or even constituted, with minimal risk of injury" (Smuts and Watanabe 1990, 160; see also Whitham and Maestripieri 2003). (In that sense, it is perhaps comparable to the above example of mounting among male lions.) Diddling is a very intimate and hence dangerous activity, at least for the one who is being diddled—he puts his future reproductive success (and a lot of his future pleasure) in the palm of a potential rival's hands. It should not surprise us, then, that diddling is one of the less common greeting behaviors in baboons, especially among young adult males. Then again, because of its potential costs, diddling provides information that is more reliable than any other kind of behavior, its intimacy being a sure sign of truthfulness.

Diddling is an interesting case study because it invites us to think about how to define our own concepts of homosexuality and sexuality. Some scientists claim or assume that diddling is an example of homosexual behavior (see, e.g., Smuts and Watanabe 1990, 148, 169; Bagemihl 1999, 18–20, 74–75); others claim or assume it isn't (see, e.g., Colmenares, Hofer, and East 2000). Unfortunately, neither position provides any real arguments for their claim or assumption. How can they make their case?

In our general introduction, we interpreted the term "homosexuality" to refer to any kind of sexual state or activity, whether physical or mental, which involves or focuses on one or more individuals of the same sex. To be sure, diddling does involve two individuals of the same sex, but is it a *sexual* activity? Given the variety of views on the nature of sexuality, the answer thus depends on which of these views one subscribes to.

According to the genital view, diddling is a sexual, and thus a homosexual, behavior, for the genital view includes as sexual all genital contact, regardless of the context in which it occurs. According to the arousal view, diddling is not a homosexual behavior, as it does not seem to be associated with any kind of physical sexual arousal, such as an erection or ejaculation, in any of the involved partners. Finally, according to the pleasure view of sexuality, it is unclear whether diddling is a homosexual behavior, because we don't know whether baboons experience pleasure when touching or holding each other's penises, even though it seems they enjoy it less than actual mating.

Animal scientists, however, do not appeal to any of these intrinsic views or arguments when discussing the nature of sexuality and homosexuality. In fact, they often appeal to the reproduction variant of the extrinsic view, which defines sexuality as any activity or state, whether physical or mental, which occurs in the context of reproduction. Nathan Bailey and Marlene Zuk (2009, 441), for example, define homosexual behavior as "interactions between same-sex individuals that also occur between opposite-sex individuals in the context of reproduction." This definition excludes a number of behaviors that only occur in same-sex interactions, which seems to be the case for diddling.[20] Greeting behaviors also occur among female baboons and between male and female baboons, but these behaviors appear to formally differ from diddling. For instance, male-female greetings in olive baboons never involve genital touching (Smuts 1985; Colmenares, Hofer, and East 2000). According to the reproduction view, diddling is not homosexual behavior because it is not sexual behavior in the first place, given that such behavior does not occur between male and female baboons "in the context of reproduction" (whatever that may mean).

Interestingly, Bagemihl also subscribes to the reproduction view, yet he himself considers diddling as an instance of homosexuality. It seems, however, that he is inconsistent in applying his definitions of sexuality and homosexuality to this case. In defining homosexuality, he argues that one should "take comparable behaviors *in the same or closely related species* as the point of reference: any activity between two animals of the same sex that involves behaviors independently recognized (usually in heterosexual contexts) as courtship, sexual, pair-bonding, or parenting activities is classified as 'homosexual'" (Bagemihl 1999, 98–99; italics in original). Bagemihl then proposes a "'broad' definition of homosexuality" that closely resembles Bailey and Zuk's: "*Homosexuality*, as the term is used in this book, refers not only to overt sexual behavior between animals of the same sex, but also to related activities that are more typically associated with a heterosexual or breeding context" (99; italics in original). Diddling is not an overtly sexual behavior, in that it is not associated with physical sexual arousal, for example, and neither is it typically associated with heterosexual or reproductive behavior. By his own definition of homosexuality, then, Bagemihl should not consider diddling as homosexual.

An important issue with appealing to the reproduction view of sexuality is that it can be seen as yet another manifestation of what Bagemihl himself repeatedly denounces as "a heterocentric view of animal behavior" (95). Why should we consider heterosexual interactions in animals as a point of reference in defining homosexual interactions? Surely, both Bagemihl and Bailey and Zuk would agree with the prima facie objection that sex cannot be reduced to its reproductive function and hence should not be defined functionally. At least in humans, for example, masturbation is clearly sexual, and it certainly should still be seen as sexual even if no men and women ever mutually masturbated. Bagemihl charged his colleagues in the animal sciences with attempting to desexualize animal homosexuality, but he himself barely fares better. In some parts of *Biological Exuberance*, he clings to the ideologically old-fashioned and narrow reproduction view of the nature of sexuality. In other parts, he chooses to employ an extremely broad definition of the sexual nature of homosexuality—a definition that even allows him to include parenting as a homosexual behavior (for a similar critique, see Poiani 2010, 11),[21] considering parenting as sexual because it is functionally linked to reproduction and reproductive success.

In the end, Bagemihl is not the only one who is confused about the nature of sexuality. The question is indeed whether we will ever be able to provide a philosophical definition of the concept of sexuality that enables us to both capture all practices and mental states that are currently or commonly considered sexual and exclude all practices and mental states that are currently or commonly considered nonsexual. As our brief review has illustrated, existing definitions all struggle with more or less fundamental problems. If these attempts already fall short of explaining the sexual nature of some prototypical sexual activities, why would we expect conceptual analyses to tell us anything about whether borderline cases like diddling are sexual? (See also Soble 2008; Christina 2017.)

This conclusion may seem pessimistic, but it need not be a major problem. Of course, it would be convenient to have a crystal-clear definition of sexuality. In the case of diddling, for example, it would help us determine whether diddling deserves to be discussed in a handbook on animal homosexuality. But if the final word is simply that diddling is and always will be a borderline case, this is still helpful, if only because it compels biologists to be careful in using terms like "sexual" and "homosexual" and to elaborate on their usage and ambiguities.

"ISOSEXUALITY" AND ALL THAT

We return now to the question we opened this chapter with: Is there such a thing as animal homosexuality? In the previous sections, we argued that at least some animal homosexual behaviors are more than mere reflex phenomena, and that they resemble human homosexual behaviors through their association with homosexual mental states, such as desires and preferences. Still, some skeptics believe that the differences between animal and human homosexuality dwarf their similarities, and therefore they object to the use of the term "homosexuality" in the context of the animal sciences. To avoid confusion, wouldn't it be better to search for alternative terms?

Zoologists are not the only scientists to struggle with this terminology question. It has also been debated in the contemporary anthropological and historical literatures on homosexuality, where some scientists are equally reluctant to label premodern or non-Western male sexual behaviors, desires, and preferences as "homosexual." In a foreword to the second edition of his *Guardians of the Flutes*, for example, anthropologist Gilbert Herdt noted that he had erred in the book's first edition when describing and interpreting certain sexual practices among the Sambia people of Papua New Guinea as instances of homosexuality. Boys and adolescents of this clan can only obtain manhood during a special initiation ceremony in which they are asked to swallow the semen of an older clan member (see also chapter 3). Is this fellatio an instance of homosexuality? At first, Herdt thought it was. Later on, however, he changed his mind. Given the historical baggage that, in his view, is attached to the term "homosexuality," he preferred an alternative: "It is no longer useful to think of the Sambia as engaging in 'homosexuality,' because of the confusing meanings of this concept and their intellectual bias in the Western history of sexuality. I would prefer to think of their practice as belonging to a larger genus of same-sex practices that might be called 'Boy-Inseminating Practices'" (Herdt 1994, xiv; see also Halperin 2002).

The argumentation behind such proposed changes to terminology is straightforward. The term "homosexuality" was originally invented to refer to a very specific set of sexual behaviors, desires, preferences, and identities that occurred primarily, if not exclusively, in industrial and postindustrial Western

European and North American societies. In our general introduction, we named this phenomenon "modern homosexuality" (see also chapters 1 and 3). The Melanesian sexual practices described by Herdt are fundamentally different from this modern variant of homosexuality. Ergo, the term "homosexuality" cannot apply to the sexual behavior of the Sambia people.

The problem further escalates when we consider sexuality in nonhuman animals. Some would say that the differences between animal and human homosexuality are much greater than the already substantial differences between Western and non-Western kinds of homosexuality, including Herdt's "boy-inseminating practices." It should not surprise us, then, that some scientists and philosophers object to using the term "homosexuality" and its derivations, such as "gay," in zoological contexts.

A common objection to the use of "human terms" in the animal sciences is its risk of anthropomorphism (Fausto-Sterling 2000). In previous sections, we distinguished between two variants of anthropomorphism. Categorical anthropomorphism holds that it would be a category mistake to ascribe homosexual mental states to nonhuman animals because they are not capable of having the mental states that are usually associated with human homosexuality. We objected to this claim on both conceptual and empirical grounds, arguing that there are in fact such things as homosexual desire, preference, and orientation in the animal kingdom. We also objected to the critique of situational anthropomorphism and argued that it is perfectly avoidable by anthropomorphizing more carefully. What else could be wrong with "humanizing" animal homosexuality?

The following examples can perhaps clarify what critics find objectionable or problematic about anthropomorphism. Some zoologists believe that homosexual behavior among juveniles helps them practice sexual skills, such as motor and social skills, that may prove helpful in heterosexual contacts throughout adult life. The sexual practice hypothesis has in fact received empirical support in many mammal species, ranging from rodents and ungulates to dolphins and primates (for an overview, see Poiani 2010). According to Roughgarden (2009, 139), however, this hypothesis is just another attempt to explain away homosexual behavior in animals: "This argument has been a familiar escape over the years from dealing with the reality of gay animals—gayness in youth is necessary for straight life later on. Really, though, let's be honest. Homosexual sheep and cattle are actually gay, not playing make-believe."

Literary theorist Susan McHugh, however, takes offense at Roughgarden's usage, if only because it detracts from the singularity of the sexual life of animals. Sheep are just sheep, she argues: "Rams identified by scientists as 'straight' prefer the company of males, self-segregate from females, and regularly engage in anal-penetrative sex with each other except when presented with a ewe in season and so, concludes Roughgarden, 'are actually gay.' Or are they all just sheep? Queer theorists argue against this sort of 'bed check' approach that equates behaviors with identities among people, in part because it enables the same old identity categories to cover over wildly divergent relations and activities" (McHugh 2011, 158).

McHugh's argument is related to Herdt's against the use of the word "homosexual" in a cross-cultural anthropological context. "Gay" refers to a sexual identity category in (a particular subset of) humans, and human sex life is "wildly divergent" from the sex life of animals, so the word "gay" should not be used to refer to animal behaviors.

A similar sentiment can be found in an essay about penguin sexuality written by biologist Marlene Zuk, an example that is frequently turned to in debates about human sexuality. The gay movement often refers to the many male homosexual penguin couples in zoos as evidence for the normality of human homosexual parenting, while some conservatives pay tribute to the dedication of heterosexual male penguins in protecting their nest and offspring. However, Zuk argues that these ideological debates cannot be settled in the animal park. Moreover, they divert our attention from what is potentially unique in the secret sex life of penguin species: "If we use animals as poster children for ideology, . . . we risk losing sight of what is truly interesting and important about their behavior" (Zuk 2006, 917). By anthropomorphizing animal homosexuality, such as when we use the term "homosexuality" for these phenomena, we throw away our chance to gain real insight into the unique determinants of their behavior.[22]

A related concern about anthropomorphizing animal homosexuality is that it may contaminate the animal world with ingrained prejudices about human homosexuality, which may prevent further zoological research. While the *Melolontha* story can help mitigate this concern, Bagemihl provides a potentially worrying example. He mentions a preliminary scientific report on killer whale behavior, written in 1979, that explicitly discussed homosexual activities among male killer whales (*Orcinus orca*). Later, the entire report was repeated verbatim in a government document for the US Marine Mammal Commission, with the sole exception of the passage on homosexuality, which had been deleted. Bagemihl (1999, 105) concludes: "It is no surprise, then, that many scientists . . . continue to harbor the erroneous impression that homosexuality does not exist in animals or is at best an isolated and anomalous phenomenon."

Finally, scientists worry about humanizing animal sexuality because it may thwart their own work and open the door for sensationalism in scientific research. In her analysis of the use of human sexual terms in animal sociobiology, zoologist Patricia Gowaty mentions a scientific paper entitled "Homosexual Rape in Acantocephalan Worms." In Gowaty's view, the paper was greeted with "a lurid snicker" and subsequently cast aside because of its perceived sensationalism, which the audience "felt represented a tactic akin to a Madison Avenue sales pitch" (Gowaty 1982, 631). Either way, the paper wasn't read seriously.

Overall, we agree that there are some good reasons to avoid carelessly using human sex terms in the animal sciences, to the extent that such carelessness may result in an erroneous understanding of both animal and human homosexuality. Therefore, some scientists and philosophers have suggested to use new sexual terms that are, in Gowaty's words, more "descriptive, adaptable, operational, and unemotional" (Gowaty 1982, 631). (Here the reader may be reminded of a similar terminological debate discussed in our general introduction.) Gowaty herself, for example, suggested that the term "homosexual" be replaced with "unisexual" in the study of nonhuman animals; others proposed similar alternatives, such as "isosexual," "intrasexual," or "ambisexual" (for a list of such terms, see Bagemihl 1999, 97). The neutrality of these alternatives, however, is both their main advantage and disadvantage: they have little or no meaning in colloquial language. This may promote communication among scientists within one domain, but it also hampers science popularization and communication across disciplines. In translating their findings to a general audience, scientists must ultimately appeal to colloquial terms, including "homosexuality." Of course, this is a minor issue, given that science popularization often uses different terms than the science it popularizes because comprehensibility requires so and is thought to compensate for accuracy.

Still, there are additional disadvantages in using alternative terms. Most of the proposed terms, such as "isosexual" and "intrasexual," originate in the late nineteenth-century sexological literature on human homosexuality, in which they denote specific subgroups of human homosexuals as defined by contemporary science. This legacy could cause confusion. Moreover, some of these terms have since been given new meanings in the informal language of certain subcultures. In some contemporary gay circles, for example, "ambisexuality" refers to a subset of bisexuals who do not seem to be bothered by the heteronormative expectations of contemporary society. Finally, alternative terms may convey the impression that all kinds of animal homosexuality are fundamentally similar—a suggestion that is loudly contradicted by nearly all specialists in the field. As Bagemihl (1999, 80) puts it: "Homosexuality . . . is not a uniform phenomenon in either animals or people: it takes many forms, and it exhibits numerous variations and idiosyncrasies." Aldo Poiani (2010, 34) concurs: "Although human sexuality may well be different from sexuality in other species, I have no reason to believe that sexuality in all the other vertebrates, or even among the other primates, represents a unitary phenomenon, with humans being the only exception."

A last objection to the use of alternative terms for animal homosexuality is that they suggest that there are no important similarities between some kinds of animal homosexuality and some kinds of human homosexuality, thus increasing the risk of missing important information when studying both phenomena in comparative analyses. It should also be noted that some scientists appear to have a double standard in this context: while they hesitate to ascribe homosexuality to animals, they do not hesitate to ascribe to animals other behaviors, mental states, or capacities, such as hunger, anxiety, or memory. According to Bagemihl, the harmful effects of anthropomorphism are given more attention in research on animal homosexuality compared to research on animal heterosexuality: "When it comes to heterosexual activities . . . , scientists are not at all adverse to making analogies with human behaviors" (Bagemihl 1999, 97).

In fact, one could argue that there are important advantages in continuing to use the term "homosexuality" in the animal sciences. The main advantage, according to Poiani, is that it "allows us to put all species on equal terms from a methodological perspective, and study them in comparative analyses that will detect differences and similarities in the association of same-sex mounting with various other variables and contexts" (Poiani 2010, 35). In our view, it would be regrettable to have our terminology, rather than our best empirical evidence, decide whether there are any similarities between human and nonhuman homosexuality. Therefore, we propose to continue using the term "homosexuality," both in the human sciences and in the animal sciences, on the understanding that it does not refer exclusively to modern homosexuality but instead refers to any kind of sexual state or activity, whether physical or mental, that involves or focuses on one or more persons of the same sex.

ANIMAL MODELS AND HUMAN HOMOSEXUALITY

The perils and possibilities of anthropomorphism chiefly concern scientists who are interested in animal homosexuality with an eye toward human homosexuality. This connection has always been an important motivation to study animal homosexuality. One of the interesting conclusions of the *Melolontha* story, for example, is that many of its protagonists could not help but compare animal homosexuality to human homosexuality. It shows how difficult it is to refrain from asking the question what, if anything, animal behavior can tell us about human behavior—even in a debate as exotic as an entomological discussion about male-male cockchafer copulation.

However, there are many reasons to conduct research into animal homosexuality. Some of the protagonists in the *Melolontha* story never once mentioned human homosexuality. Their main motivation was to understand the peculiarities of nonreproductive sex in specific insect species. The intrinsic value of knowledge sufficed for them, and a similar intrinsic interest in animal homosexuality still typifies a large part of current scientific research on the topic.

Another portion of this research stems from economic considerations, such as in the financial optimization of sheep farming. We mentioned before that a certain percentage of range-bred rams consistently prefer to court and mate with males even if they have access to females. According to some estimates, no less than 6 and up to 10 percent of the domesticated ram population have a homosexual orientation (Roselli, Resko, and Stormshak 2002). In wool-growing vernacular, these rams are known as "duds" or "dud studs" (Roughgarden 2009, 139). Sheep farmers are not particularly keen on having such animals, given that the average purchase price of a breeding ram fluctuates between three hundred and three thousand dollars and that each of these rams is expected to mate with at least fifty and up to one hundred ewes every fall. For sheep farmers, a dud stud is a drain, so it is hardly surprising to find that homosexuality in rams has been intensively researched in the animal sciences, both in the lab and in the field (Terry 2000).

Even though a great deal of research on animal homosexuality initially stems from interests outside of human homosexuality, the *Melolontha* story and even the literature on dud studs reveal the irresistible allure of seeking similarities between both phenomena. One example is a representative newspaper report on Jim Fitzgerald's endocrinological work on homosexuality in rams, which confusingly compares Fitzgerald to Sigmund Freud, noting "he's convinced their [the rams'] mothers are somehow responsible. It may go all the way back to the womb and the 'hormonal environment' of early pregnancy" (Anonymous 1991). Researchers and reporters alike seem to have difficulties with not remarking on various similarities between human and nonhuman homosexuality. These resemblances have prompted scientists from different disciplines to move in similar directions.

Research on human homosexuality, on the one hand, is replete with references to the animal sciences. This is particularly true of evolutionary research on human homosexuality, as we explain in the next chapter, but it also applies, for example, to work on the neuroanatomy of homosexuality. The title of Simon LeVay's (2011) editorial "From Mice to Men: Biological Factors in the Development of Sexuality" reflects his source of inspiration.

Animal scientists, on the other hand, often draw attention to the potential relevance of animal homosexuality for research into human homosexuality. In a rare interview about his magnum opus on animal homosexuality, Bruce Bagemihl underlines that "the implications for humans are enormous" (Harrold 1999).[23] The use of animal models is run-of-the-mill in experimental research, particularly in medicine. In the past, these models have generated new insights into the causes and treatment of numerous human diseases, ranging from epilepsy and cancer to glaucoma. In psychology and in the behavioral sciences, animal models have often proved their use. One wonders, then, if animal models can help us unravel some aspects of human homosexuality. The following is an attempt to answer that question.

It is important to first consider why, in research on human homosexuality, we would use animal models at all. First, there are moral reasons. Many experimental manipulations in which human sexual behaviors or desires are a dependent variable will never be approved by scientific research ethics committees. Experimental research on animals continues to be controversial, but from a moral point of view, many consider it less problematic than comparable experiments on humans. Second, some animal models have already proved their use in medical research on human sexuality. Rat research, for example, has contributed to the development of medical treatments for erectile dysfunction. Of course, the usefulness of animal models in learning about erectile dysfunctions does not entail that such models will also help us understand more cognitively and socially rich phenomena, such as desires and preferences. Third, there are pragmatic and economic considerations. Lab animals are chosen in part because they are easy to breed and manipulate. Moreover, they are relatively small and harmless to humans, and they have a comparatively short life span, which means that the full development of an individual can be studied in relatively short-term research. In the past five decades, such research has generated a huge amount of data about various animal species, including fruit flies and rodents, and it would be a shame not to take advantage of animal models in attempting to unravel the mysteries of human homosexuality.

Not all animal species are good model organisms to better understand aspects of human homosexuality. The nineteenth-century criminologist Alexandre Lacassagne (1882, 7) noted that insects, for example, cannot help us understand criminal behavior in humans, which in his view included homosexuality, as they are far too distant in the phylogenetic tree (and too physiologically different) to share any interesting analogies with humans. So how do scientists go about selecting relevant animal models? The three reasons mentioned above can serve as guidelines here. First, is it morally acceptable to use this particular species? (Or, more cynically: Will the ethics committee object to my using this species?) Second, what do we already know about the sexual physiology and behavior of this species? Third, to what extent have earlier studies on this species proved relevant for the study of human sexuality? Fruit flies easily fulfill the first two criteria, but until now, their sex life has not been very informative in research on human sexuality.

Rats, by contrast, seem to be better candidates for model organisms. They exhibit homosexual behaviors, desires, and even preferences, although there is some debate about the exclusivity and stability of these preferences (Coria-Avila 2012; Triana-Del Rio et al. 2015). Furthermore, compared to fruit flies, rats are much closer to humans from a phylogenetic and from an ecological point of view. Moreover, in the past seven decades there has been a great amount of research on the (neuro)endocrinological aspects of homosexuality in rats. Already in the 1940s and 1950s, biologist Frank Beach revealed a correlation between certain kinds of homosexual behavior in rats and prenatal exposure to female sex hormones (Beach 1948). Beach subjected male rat fetuses to high doses of estrogen in utero and found that, in later life, they displayed a conspicuous tendency toward lordosis behavior, that is, they frequently adopted a posture typically taken by females during heterosexual copulations. The primary characteristics of lordosis are immobility, a lowering of the forelimbs, an extension of the hind limbs, and a ventral arching of the spine. In the past, many scientists, including Beach, believed lordosis behavior was a plain invitation to copulate and, when observed in male rats, was homosexual behavior. Contemporary researchers agree that lordosis facilitates penetration, but they consider it a reaction to a mounting male, rather than an invitation to such mounting (Marson and Wesselmann 2017).

Still, later experiments confirmed some of the results of Beach's pioneering work, showing that neonatal endocrine manipulations cause the sexual preference of some male rats to change from females to other males (Bakker et al. 1993; Olvera-Hernández, Chavira, and Fernández-Guasti 2015). In fact, quite a few findings from contemporary neuroendocrinology, which studies the impact of hormones on the developing brain, do fit the predictions derived from animal models, although with some caveats. For example, human female fetuses who have been exposed in utero to high doses of male sex hormones, such as testosterone, are more likely to develop a homosexual orientation (Balthazart 2011a, 2011b). Surprisingly, the same is true for some men: high doses of testosterone during crucial periods of embryonic development increase the likelihood of their becoming homosexual. This finding is consistent with the fact that some homosexual men display a number of hypermasculine characteristics, such as a larger than average penis (Bogaert and Hershberger 1999), and recent rodent models show similar results. Male rat fetuses exposed to both low and high levels of androgens are all more likely to develop a homosexual preference.

Moreover, this somewhat paradoxical effect of androgens in rats may be instrumental for our understanding of individual differences in human male homosexuality, as one recent review suggests:

> When androgen signaling is increased in male rodents . . . , males continue to exhibit male-typical sexual behaviors, but their sexual preferences are altered such that their interest in same-sex partners is increased. Analogous to this rodent literature, recent findings indicate that high level androgen exposure may contribute to the sexual orientation of a subset of gay men who prefer insertive anal sex and report more male-typical gender traits, whereas gay men who prefer receptive anal sex, and who on average report more gender nonconformity, present with biomarkers suggestive of low androgen exposure. (Swift-Gallant 2019, 23)

The reader may doubt whether the gay sciences should (re)naturalize the distinction between "passive" and "active" homosexuals (and we definitely share some of these doubts), but these findings are nevertheless intriguing and highlight the potential of animal models for the study of human homosexuality.

Be that as it may, experimental studies on homosexuality in rats have failed to impress some philosophers of science. These philosophers have two related worries.

A first concern is that the behavior in the model organism is only superficially similar to the target behavior in humans. Anne Fausto-Sterling (1995), for example, argues against using male lordosis-rats as model organisms for human male homosexuality, given that lordosis is absent in humans and the penetrating partners' behavior seems as homosexual as the behavior of the lordosis-rats, at least according to the definition of homosexuality we provided in our general introduction (see also Stein 1999).[24] To some extent, Fausto-Sterling's criticism is pertinent.[25] If we really want to know what homosexuality in rats can teach us about human homosexuality, then the focus on male lordosis-rats surely seems biased and probably not all that informative (Marson and Wesselmann 2017). Fausto-Sterling is also right that there was (and perhaps still is) a notable tendency among sex scientists to overestimate the importance of the association between human male homosexuality and femininity. Lordosis is a behavior associated with adult female rats, but homosexuality concerns the sex of the subjects toward which sexual behavior is displayed and should not be confused with the display of behaviors typical of a different sex.

The question is, however, whether such issues are enough to warrant Fausto-Sterling's wholesale skepticism of the value of animal models in studying human homosexuality. In the case of rat research, her reserve seems exaggerated. If, for example, it is true that "passive" homosexual male rats differ endocrinologically from both "active" homosexual and heterosexual male rats, and if we are interested in the role of hormones in the development of homosexuality in humans, then it seems obvious to focus first on the "passive" partner. Second, researchers are now quite careful in distinguishing between feminization and homosexuality in male rats (see, e.g., Teodorov et al. 2002). It is simply not true anymore that "homosexuality in rats is equated with lordosis" (Stein 1999, 173). Most crucially, however, there is very suggestive evidence that neural and endocrine mechanisms produce variation in rodent sexual behavior, and this evidence has inspired research that has shown that variation in human sexual orientation is partly explained by variation in exposure to certain sex hormones in utero (Balthazart 2011a, 2011b).[26]

A second concern about animal models of human homosexuality is that human homosexuality is much more complex than animal homosexuality (Stein 1999). According to critics, animal models of homosexuality force us to face the fact that human homosexuality is an extremely complex and heterogeneous phenomenon. This is true not only on the level of its dimensions or manifestations, as we explained in the general introduction, but also on the level of its underlying causal mechanisms. The heterogeneity of homosexuality is one of the main obstacles for the use of animal models in studying human homosexuality. If we cannot extend the few reliable and robust research results in one specific subgroup of homosexual men to other subgroups, then why would we expect to be able to extend the results of rat research to human homosexuals? If relevant findings are not even universally valid within one species, then some may argue that it is even more implausible that they would be universally valid across different species. It is indeed tempting to overlook the dissimilarities in homosexuality between rats and humans. In humans, homosexual behavior is often an expression of a homosexual orientation, and we know that such orientation is rare in noncaptive populations of other animal species. Therefore, many animal homosexual behaviors will require a different explanation. We may have criticized the casualness with which the error hypothesis has been used in the context of animal homosexuality, but we also acknowledged that it likely does explain many animal homosexual behaviors. In human homosexuality, by contrast, error plays a more minor role.

Furthermore, many recent findings suggest that the neuroendocrine mechanisms involved in homosexuality differ in humans and in rats, and this dynamic seems to be more complex in humans: "Clinical observations support the hypothesis that in human prenatal development, sexual brain differentiation is subject to effects of androgens, but these are not of the hormonal-robot type found in subprimate mammals, in which sex steroids, in the set of behaviors studied, typically exert a simple on–off effect on sexual functioning, both in their organizational and activational effects . . . , and there are certainly other unidentified factors that modulate or override androgen effects on the central nervous system" (Gooren 2006, 593).

Exposure in utero to a lack or surplus of sex hormones does not determine human sexual orientation. Interestingly, less than half the individuals who experienced serious hormonal dysfunctions during embryonic development actually develop into homosexuals (Balthazart 2011a, 2011b). Most likely, hormonal exposure is not even necessary. In any case, variation in hormonal exposure only accounts for a part of the variation in human sexual orientation, and it seems reasonable to conclude that some homosexuals do not owe their sexual orientation to the presence or absence of certain sex hormones in the womb. We must look for the causes of their orientation elsewhere. Moreover, some of the hormonal effects on homosexual behavior in rats are not found in humans. In rats, for example, most effects of testosterone in the brain may be mediated via its aromatization into estrogen, and it has been theorized that this may also be the case in humans. However, a recent review of the empirical literature on this issue concludes that "for the development of human male gender identity and male heterosexuality, direct androgen action on the brain seems to be of crucial importance, and the aromatization theory as derived from experiments in rodents may be of secondary importance for sexual differentiation of the human brain" (Swaab 2004, 302).

Our conclusion is that the use of rats (and other animals) as model organisms for human homosexuality is primarily of heuristic or methodical value. This is not a negative assessment, since many researchers use animal models precisely for that reason. Biologists do not try "to produce 'models of' a phenomenon that make an enduring truth claim about how something works," but rather "generate 'models for' particular practical purposes" (Nelson 2018, 48). Moreover, it is hard to deny that experimental rat research about the correlation between hormones and homosexuality has marked the contours of a potentially fruitful line of research on human homosexuality.

Therefore, we disagree with those philosophers who argue that animal models of human homosexuality are entirely useless. Animal models are exceedingly useful scientific instruments to generate hypotheses about the nature and causes of a wide variety of human peculiarities. Importantly, the fact that human behavior tends to be more complex than the behavior of the model organisms would not surprise researchers who use model organisms.

These models are used with the hope that "data and theories generated through use of the model will be applicable to other organisms, particularly those that are in some way more complex than the original model, especially humans" (Ankeny and Leonelli 2020, 2). Therefore, researchers expect that there will be differences but hope at the same time that "the overarching themes will translate" (Swift-Gallant 2019, 28), given the genetic similarities between the two species and the phylogenic, physiological, cognitive, and ecological proximity of the model to the target species. Surprising findings about rodent sexuality will sometimes lead to rather spectacular new insights into human homosexuality, such as when the observed paradoxical effect of androgens on rodent homosexual behavior shed new light on the biological heterogeneity of human (male) homosexuality. However, findings about rodent sexuality may also fail to generalize to human sexuality, but this is not necessarily a major problem, since negative and null results are still scientifically valuable. Learning that something is not the case also constitutes scientific progress (Mlinarić, Horvat, and Smolčić 2017). So even if the predictive value of animal models is often more limited than assumed (Shanks, Greek, and Greek 2009), this does not mean that their use in the biomedical and behavioral sciences should be opposed.

CONCLUSION

Is there such a thing as animal homosexuality? The question deserves a nuanced answer. All too often, zoologists have tried to explain away homosexual behaviors in animals, dismissing them as not *really* homosexual or even sexual at all, despite their conspicuous similarities with human homosexual behaviors. If anything, these similarities suggest it is possible to use animal models in studying human homosexuality. Nevertheless, there are good reasons to be careful in describing and explaining animal homosexuality. In this chapter, we discussed some possible pitfalls, but we feel that they are outweighed by the possible profits of doing proper research on animal homosexuality.

Much of this chapter revolved around one meaning of homosexuality's supposed naturalness, that is, its abundance in so many parts of the animal kingdom. Other meanings of "naturalness" will soon surface in the next chapters. In the following chapter, we discuss homosexuality's naturalness by asking whether it is or has been the result of natural selection. In chapter 4, we ask whether and why naturalness matters to homosexuality's former status as a disorder. Finally, the epilogue details the complex relationship between naturalness and morality: If homosexuality is natural, then what, if anything, does that mean for our moral attitudes toward homosexuals?

[CHAPTER 3]

Beyond the Paradox

On Homosexuality and Evolutionary Theory

Si la race se perpétue, les invertis n'y sont pas pour grand'chose.

CHARLES FÉRÉ

Most of our assumptions have outlived their uselessness.

MARSHALL MCLUHAN

Alternately labeled as a puzzle (Berman 2003), a paradox (Camperio-Ciani, Corna, and Capiluppi 2004, 2217), a mystery (Rahman and Wilson 2003, 1357) or a conundrum (Peters 2006), homosexuality remains one of the showpieces in the evolutionists' curiosity cabinet. If it is true that human male homosexuality goes back beyond the beginning of humanity, if genes contribute to homosexuality's development, and if homosexuals produce little or no offspring, then why does it still exist? Charles Darwin once said that "natural selection is daily and hourly scrutinising, throughout the world, every variation, even the slightest; rejecting that which is bad, preserving and adding up all that is good" (Darwin 2008, 66). Given its impact on reproductive success, homosexuality would seem to be a paragon of a "bad" variation. So how did it manage to elude natural selection?

We begin this chapter by mapping out the history of this paradox, following its origins all the way to a cautious first formulation (and even more cautious first solution) by the ecologist George Evelyn Hutchinson in the late 1950s. Hutchinson's work turned out to be the first contribution to what has now become a cottage industry of evolutionary hypotheses of human male homosexuality. We then focus on two hypotheses that have enjoyed some scholarly popularity and empirical support: the "sickle cell" hypothesis and the "sugar daddy" hypothesis. Targeting these same hypotheses, we question two of their central assumptions: a claim about the homogeneity of human male homosexuality and a claim about its impact on reproductive success. Our main criticism is that some contemporary evolutionary scientists ignore the diversity of homosexual phenotypes or forms of homosexuality, particularly by overestimating the importance of the form that we referred to as modern homosexuality in our general introduction, which we will discuss in more detail here.

However, readers should not take our criticism to mean that evolutionary theory is irrelevant in understanding human male homosexuality. The main argument of this chapter is that different forms of homosexuality may require different evolutionary hypotheses. In the final sections of this chapter, we detail two contemporary evolutionary perspectives on the topic, the alliance formation hypothesis and the kin influence hypothesis, both of which go some way to account for the diversity of homosexuality.

One way to read this chapter is to consider it as an illustration of two of the uses of the philosophy of science that we discussed in the general introduction. First, we critically engage with some of the assumptions underlying prominent evolutionary explanations of human male homosexuality. Second, we set up a dialogue between the life sciences (genetics, evolutionary biology, and endocrinology) and the human sciences (sex history, anthropology, and sociology). In doing so, we hope to contribute to new ways of conducting evolutionary research on homosexuality.

A PARADOX AND ITS HISTORY

It is perhaps unsurprising to learn that it was not Darwin who first considered homosexuality an evolutionary oddity. In fact, he never once mentioned any variant of its name in either his oeuvre or his thousands of letters. When one of his correspondents, the British diplomat Robert Swinhoe, presumed to discuss homosexual behavior and homophobia among Chinese dogs—"a curious story of Dog-morality which may be of some use to you" (don't ask)—Darwin politely ignored the matter (Darwin, n.d.). For one thing, the topic was probably unseemly for a Victorian gentleman to discuss, although Darwin did acknowledge same-sex pairings in partridges and starlings in *The Descent of Man* (Brooks 2021).

More importantly, questions about the evolutionary persistence of homosexuality needed the so-called modern synthesis to surface. The modern synthesis was a double set of theoretical developments in evolutionary theory that came about in the first half of the twentieth century (Grene and Depew 2004). The first synthesis joined together Darwinian thinking and Mendelian genetics into a new discipline called population genetics (1920–30), while the second unified population genetics with a number of other emerging biological disciplines (1930–50). By introducing a whole range of new concepts and arguments and, most importantly, a mathematical theory of selection in natural populations, the modern synthesis provided significant raw material for the paradox of homosexuality and natural selection. In addition, it outlined some intriguing solutions. One of the many innovations of the modern synthesis was Theodosius Dobzhansky's theory of balancing selection, which we argue has played an important role in the development of evolutionary thinking about homosexuality.

Around the same time as Dobzhansky's work in the mid-1900s, Franz Kallmann was conducting his pioneering twin studies of the genetics of homosexuality, which we discussed in chapter 1. The reader will recall that these studies favored Kallmann's own view that homosexuality is "a genetically controlled imperfection" (Kallmann 1952a, 283), but they also revealed that male homosexuals produce little or no offspring. Of the eighty-five subjects in his first twin study on male homosexuality, only eleven were married: "Most of these marriages lasted no longer than a few months, and only three of them were fertile, resulting in a total fertility quota of five children, three boys and two girls. Unfortunately, in no instance has it been possible to confirm the paternity of the legal fathers beyond reasonable doubt" (287).

Much like the subjects he studied with schizophrenia, Kallmann's homosexual twins married less frequently, tended to remain childless when married, and produced fewer children, if any, when compared to heterosexual controls (Kallmann 1953, 133). Interestingly, Kallmann's work shows, on numerous occasions, a distinct familiarity with theoretical developments in evolutionary biology, such as when he emphasizes the selective disadvantage of various psychiatric illnesses. For some reason, however, he never asked why natural selection had not been able to eliminate these "genetically controlled imperfections."

Some years later, Kallmann's homosexual twin studies landed in the hands of ecologist and polymath George Evelyn Hutchinson (see fig. 3.1). Exactly a century after the publication of Darwin's *On the Origin of Species*, he first brought together all the pieces of the paradox in a 1959 paper published in the *American Naturalist* and suggested a tentative solution. In his autobiography, Hutchinson (1979, 128) jokingly remarked that this paper made him "an unwitting very minor founding father of the Gay Liberation Movement."

How did Hutchinson come to take an interest in the evolution of homosexuality? First, it was an open secret that homosexuality was common practice among some of the early interwar Oxbridge student population, when Hutchinson was a student there. Perhaps that helped stir his interest in the topic, as it did, for example, for the philosopher George Edward Moore.[1] Second, and more generally, people's attitudes toward sex had changed since the end of the Victorian era. Even though in the 1950s many

Figure 3.1: George Evelyn Hutchinson with baby potto during his final class at Yale University (1971). © William K. Sacco.

still considered it a crime or, increasingly, a mental disorder, male homosexuality had become a debatable topic in the Anglo-Saxon world. This change was in part due to the biostatistical work of Alfred Kinsey, Frank Beach,[2] and various other sex researchers who dealt with both male and female homosexuality in the late 1940s and early 1950s. (For more on their work, see chapters 2 and 4.) Finally, Hutchinson was an intellectual gourmand with an appetite for topics that were deemed peripheral or even marginal both in his own specialty of ecology and also in the wide variety of other sciences that interested him (Slack 2010). During the 1940s and 1950s, he mused on such topics in a column in *American Scientist*, appropriately named Marginalia, including, among many other things, rhabdomancy, climate change, gothic architecture, and the importance of ornithology.

On the final page of one of his last columns, Hutchinson (1957, 95) made a very cautious remark about "the whole question of the direction of sexuality," echoing his psychoanalyst contemporaries' firm conviction that "there is apparently absolutely no evidence that there is any genetic basis for sexual inversion." Clearly, he had not yet seen Kallmann's studies, but he anticipated them with a very tentative proposal based on Dobzhansky's ideas about balancing selection: "If there were [any genetic basis for sexual inversion], it would obviously be selected against and would have disappeared *unless a balanced polymorphism were involved*" (italics ours).

Two years later, Hutchinson (1959) finally hazarded to sort out the topic. His analysis is as simple as it is elegant. First, he referred to Kallmann's work as supporting "the commonsense view that homosexuality in western cultures is likely to be correlated with a low fertility" (82). Second, he noted he was now inclined "to accept the findings provisionally as indicating at least some genetic determination, though probably not as much as would appear at first sight" (85). Hutchinson was clearly impressed by Kallmann's studies (though less so by his "somewhat arcane expressions" [88]), but at the end of his paper, he pointed out an important difference of opinion. In his view, Kallmann's distrust of psychoanalytic explanations of homosexuality was rather unfounded. By his own account, Hutchinson sought to reconcile the two disciplines, hypothesizing that the genes involved in homosexuality may play a role in the development of the neuropsychological mechanisms underlying certain infantile identification processes and object relationships, then thought by psychoanalysts to play a role in becoming gay (87–88). (Such a conciliatory project would likely raise some eyebrows among contemporary biologists.)

If Kallmann was right about both fertility and genetics, Hutchinson continued, then from an evolutionary point of view, there were three options. A first option is that homosexuality is becoming increasingly rare because of high selection rates. Hutchinson quickly dismissed this hypothesis as it "would hardly be in accord with historic evidence, and would raise the difficulty of providing an explanation as to how the genotypes involved reached their former high frequency" (86). A second option is that homosexuality-related genes are being maintained in the population as ever-new mutations. The problem with this second option, Hutchinson noted, was that other conditions caused by single genes, such as Mendelian disorders, are much more rare than (male) homosexuality, the prevalence of which had recently been estimated by Kinsey at around 10 percent.

The third and final option, favored by Hutchinson, is that homosexuality is maintained in the human population as part of some kind of "balanced polymorphism" (Hutchinson 1957, 95; 1959, 81). His basic idea was that some of the genes associated with homosexuality may be involved in the expression of traits that actually, and unlike homosexuality, boost their carriers' reproductive success. Curiously, Hutchinson never speculated about the nature of such traits. In his 1959 paper, he simply notes that balancing selection had already been proposed as a solution for various other biological puzzles, including the presumed childlessness of geniuses. Here he referred to the work of zoologist Charles Burford Goodhart, who hypothesized a genetic connection between intelligence and (low) fertility: "A possible solution to the [evolutionary] puzzle [of infertility] is that the genes responsible for lowered fertility have become associated with other qualities of high survival value, such as health and intelligence, to make up for the selective disadvantage of relatively low fertility. . . . Real genius, in the arts or sciences and even in business and politics, has often been associated with infertility; if all the best English poets, for example, had acquired hereditary titles most of them would now be extinct" (Goodhart 1957, 14).[3]

In his 1957 column, Hutchinson referred to a less controversial example of balanced polymorphism in humans: the relationship between sickle cell disease and resistance to malaria. In the next section, we argue that this relationship provided a popular template for evolutionary thinking about human male homosexuality in the second half of the twentieth century (and the first decades of the twenty-first), resulting in what we describe as various "sickle cell hypotheses."

SICKLE CELL HYPOTHESES

By the mid-twentieth century, it had become clear that many living species have several genetic variants at a number of locations in their DNA. In biology, this phenomenon became known as genetic polymorphism, and biologists were left wondering why natural selection was unable to reduce such variation, given its general tendency to do just that. One answer they came up with was that some polymorphisms are maintained in the population by various forms of balancing selection. One form, endorsed by Dobzhansky, involves a situation in which heterozygote variants have some advantage over both homozygote variants. To understand this hypothesis, known in the literature under names as varied as "heterosis," "overdominance," and "heterozygote advantage," one needs to know some basic facts about genetics.

Each of our genes consists of two variants known as alleles—one from each of our parents. For convenience, let us label them as alleles A and a. The combination of both alleles yields three possible scenarios. When alleles differ (Aa), their owner is a heterozygote. Identical alleles yield a homozygote owner, but homozygotes come in two different forms. *AA*-homozygotes are considered dominant, since the trait produced by allele *A* overrides the trait produced by allele *a*. By contrast, *aa*-homozygotes are considered recessive, meaning that the trait produced by allele *a* only occurs in homozygous individuals.

The heterozygote advantage hypothesis claims that genetic polymorphisms persist because heterozygotes enjoy evolutionarily relevant benefits that do not occur in either homozygote. In some cases, these benefits are so substantial that they compensate for the disadvantages of a harmful trait in recessively homozygous individuals. Biologists have labeled such cases "balanced (genetic) polymorphisms"—balanced because somehow the advantages and disadvantages of the various traits involved maintain a steady equilibrium. As noted before, Hutchinson hypothesized that homosexuality was part of such a polymorphism, and he wasn't the first to apply Dobzhansky's ideas. In the early 1950s, some years before Hutchinson's hunches about homosexuality, geneticist Anthony Allison (1954) famously immortalized Dobzhansky's heterozygote advantage hypothesis by discovering an interesting relationship between sickle cell disease and malaria resistance.

Sickle cell disease is a group of red blood cell disorders. In normal humans, red blood cells are elastic and disk-shaped, which allows them to move easily through small blood vessels to carry oxygen around the body. In sickle cell patients, by contrast, red blood cells are rigid and sickle-shaped. These anomalies cause them to obstruct capillaries and disturb blood supply, resulting in a variety of symptoms, including anemia, infection, and pain. In times past, many sickle cell patients died in early childhood, and even though the mortality rates for European sickle cell patients have dropped considerably in the past decades, their life expectancy is still significantly shorter than average (Maitra et al. 2017).

Epidemiology provides us with one of the most remarkable findings about sickle cell disease. As it happens, the disease occurs most often in people descending from tropical regions in the world, particularly the Middle East, India, and sub-Saharan Africa. Geneticists have explained this peculiar geographical concentration. First, they have shown that sickle cell patients are recessively homozygous (*ss*) for the β-hemoglobin locus in their DNA. Here, on the short arm of chromosome 11, *S*-alleles contain the code to produce various kinds of hemoglobin—a protein designed to bind oxygen in red blood cells. The mutated *s*-allele produces a rare and anomalous kind of hemoglobin, which, in a homozygous constellation, causes sickle cell disease. Second, and more importantly, geneticists have shown that β-hemoglobin heterozygotes (*Ss*), by contrast, do not suffer from sickle cell symptoms. While heterozygotes possess the *s*-allele, this produces only a limited quantity of anomalous, sickle-shaped red blood cells but not enough to cause any symptoms. In addition, these heterozygotes have a remarkable advantage in tropical biotopes, as their *s*-allele confers a certain amount of resistance to malaria, which is very common in these areas. The malaria parasite spends part of its life cycle in red blood cells, and sickle-shaped cells prevent the parasite from reproducing.

The relationship between sickle cell disease and malaria resistance is a rare example of a balanced genetic polymorphism. Overall, the net fitness effects of the allelic variants of the β-hemoglobin gene (*ss*, *SS*, and *Ss*) balance each other out, thus explaining why natural selection has failed to eradicate sickle cell disease. While an ideal world would only produce heterozygotes, in a world bound by the laws of genetics, a quarter of the offspring produced by two heterozygotes will still be recessively homozygous and develop sickle cell disease. This disease is thus the price a family pays for an unusual advantage conferred to some of its members: resistance to malaria.

Evolutionary scientists have always been very intrigued by the sickle cell story. In 1964, just a few years after the publication of Hutchinson's paper on homosexuality, biologist Julian Huxley teamed up with polymath Ernst Mayr and two psychiatrists, Humphry Osmond and Abram Hoffer, to write a foundational paper about the evolution of schizophrenia (Huxley et al. 1964; De Bont 2010). They hypothesized that schizophrenia might confer some hidden fitness advantages to the relatives of patients with schizophrenia, thus explaining how this heritable disorder escaped natural selection. Following the logic of the sickle cell story, they focused mainly on physiological advantages, such as resistance to infections and allergies. Later on, research shifted to psychological benefits, including intelligence and creativity. More recent research shows, for example, that artists who score high on particular components of schizotypy, a mild form of schizophrenia, also have a higher than average number of sex partners, which contradicts Goodhart's wholesale claim about the connection between genius and low fertility (Nettle and Clegg 2005).

Similarly, evolutionary scientists have followed Hutchinson's lead by applying the sickle cell logic to the evolution of homosexuality. Assuming that the reproductive success of male homosexuals is markedly lower than that of male heterosexuals, they searched for hidden advantages in relatives. The result has produced a myriad of balancing selection hypotheses about the evolution of human male homosexuality. In what follows, we discuss two examples of such hypotheses.

A first hypothesis is that female relatives of male homosexuals compensate for the latter's lowered reproductive success by producing more offspring than female relatives of male heterosexuals. Researchers have described this hypothesis as an example of sexually antagonistic selection, where the direction of natural selection on a trait differs between the sexes. The scenario has gained in popularity since the early 2000s, when Italian researchers were able to show that mothers and aunts of some male homosexuals were indeed significantly more fecund than maternal relatives of male heterosexuals (Camperio-Ciani, Corna, and Capiluppi 2004; Iemmola and Camperio-Ciani 2009). The authors also noted that the increased fecundity occurred in maternal relatives of about 20 percent of the total population of homosexual men in their study.

A second hypothesis instead focuses on male relatives of homosexual men. In the late 1990s, psychologist Jim McKnight noted that male homosexuals have much more charm and many more sex partners than (nonfamily) male heterosexuals, inferring that they must also have a bigger libido. As heterozygote relatives, the brothers and fathers of homosexual men would share these features, thus enabling them to produce more offspring than the brothers and fathers of heterosexual men (McKnight 1997). The problem with this hypothesis is twofold. First, it is based on the erroneous assumption that there is a strong correlation between the magnitude of one's libido and the number of one's sex partners. Male homosexuals may well have more sex partners simply because they are able to, rather than because of their libido. Second, and unlike the first hypothesis, McKnight did not systematically substantiate his hypothesis with relevant data.

Another variant of the second hypothesis centers on the attractiveness, rather than the fecundity, of heterosexual male relatives of male homosexuals. In the early 1990s, some researchers suggested taking into account the "physical attributes of those with homosexual genes [i.e., heterozygotes] and the ability to please their female sex partners"—attributes including "greater attractiveness" and "sensitivity" (MacIntyre and Estep 1993, 231; for a recent formulation, see Zietsch et al. 2008). The tentative idea was that a small dose of such "feminine" qualities would boost the attractiveness of heterosexual male relatives of homosexual men, much like a small dose of sickle-shaped red blood cells would boost malaria resistance in close relatives of sickle cell patients. Later studies may have contradicted this idea by indicating that heterosexual women sexually prefer markedly masculine faces. However, even more recent research indicates that women in small-scale and nonurban societies, that is, societies reflecting our ancestral environment, have a marked sexual preference for men with "feminized faces" and rounded, soft features (I. Scott et al. 2014).

The success of this second hypothesis depends on evidence about the supposed physical and psychological qualities attributed to both homosexual men and their male heterosexual relatives, as well as on further evidence about the heritability of homosexuality. Recent studies seem to favor the first hypothesis, indicating that enhanced fecundity seems to occur not in all relatives of male homosexuals but only in maternal-line women (Camperio-Ciani, Battaglia, and Zanzotto 2015), at least in Caucasian families (Rahman et al. 2008).

Collectively, however, sickle cell hypotheses suffer from a number of shortcomings. First, the initial optimism about their capacity to explain genetic variation in nature has slowly subsided in the second half of the twentieth century. By the late 1980s, scientists had discovered only six additional examples of stable polymorphisms (Endler 1986). Second, sickle cell hypotheses are designed to explain the persistence of heritability in certain traits, which means they work best for highly heritable conditions (but see Camperio-Ciani, Battaglia, and Zanzotto 2015). In chapter 1, however, we noted that heritability estimates for male homosexuality are not very high, ranging from 25 to 65 percent. Third, given the cost of generating homozygotes, sickle cell scenarios are under enormous pressure from natural selection, which implies that such scenarios are likely to be temporary evolutionary stopgaps. This implication is inconsistent with our knowledge about both the history of human homosexuality and its occurrence in nonhuman animals. Finally, and most importantly, the scope of sickle cell hypotheses is rather limited, as they focus exclusively on modern homosexuality, assuming that it is representative of all kinds of homosexuality at all times and in all places. We find this assumption very problematic, as we explain in later sections of this chapter.

THE SUGAR DADDY HYPOTHESIS

All the sickle cell hypotheses about the evolution of human male homosexuality revolve around Dobzhansky's theory of balancing selection. Another popular hypothesis about the persistence of gay genes in the population builds on kin selection theory. The biologists John Haldane and William Hamilton first developed this theory in the 1950s and 1960s. Many believe that the entomologist and sociobiologist Edward O. Wilson was the first to apply this theory to homosexuality. However, he himself once declared that the zoologist and university administrator Herman Spieth suggested it to him in a personal communication, while Robert Trivers also independently developed it in his famous paper on parent-offspring conflict (Wilson 1975, 555).

In order to understand the kin selection hypothesis, let us first reconsider the paradox we are dealing with: why do so-called gay genes persist if human male homosexuals leave little or no offspring? Generally, there are two ways of ensuring the continued existence of one's genes. A first and rather straightforward strategy is to produce offspring, since parents and children share approximately half their genes. A second, indirect strategy focuses less on direct parenthood and more on the care of close relatives instead, assuming one shares with them a smaller yet still substantial amount of genes. Some grandparents enjoy caring for their grandchildren, for example, which from an evolutionary point of view makes sense because, on average, grandparents and grandchildren still share approximately one-quarter of their DNA. By taking care of their grandchildren, grandparents also ensure the continuation of their own genes. According to the theory of kin selection, such care and commitment will decrease as the degree of kinship broadens out.

In Wilson's view, there are a number of generally recognized social roles, both in humans and in other animals, that allow individuals to indirectly pass on genes. He thus mentions a number of "distinctive [human] physiological or psychological types that recur repeatedly at predictable frequencies within societies. Some would probably be altruistic in behavior—homosexuals who perform distinctive services, celibate 'maiden aunts' who substitute as nurses, self-sacrificing and reproductively less efficient soldiers, and the like" (Wilson 1975, 311).

What services would homosexuals then provide? Wilson believes that they could act as so-called sugar daddies for their close relatives, particularly by investing time, energy, and other resources in their second-degree and third-degree relatives, like siblings or nephews and nieces. Such investments have an obvious return that boosts the likelihood of their genes' survival into the next generation—including the genes involved in developing a homosexual orientation.

Trivers provided another metaphor to capture the social role of human male homosexuals, consistent with Wilson's hypothesis, by comparing them to the biological phenomenon of "helpers at the nest," or offspring who assist their parents in raising subsequent broods or litters. Some of these helpers also build their own nest and produce their own offspring, but some give up on parenthood altogether. Examples of such helpers can be found in many bird species, including white-throated bee-eaters (*Merops albicollis*) and white-winged choughs (*Corcorax melanoramphos*). Nesting pairs of these so-called cooperative breeders have up to five helpers who are all closely related to one of the parents (Reynolds 1972). Intriguingly, Trivers added that his hypothesis allowed for the possibility that parents would induce or at least encourage such altruistic behavior in some of their offspring. That hypothesis may also work in a human context. Until well into the twentieth century, for example, many families would coax some of their members into priesthood, given the many benefits and opportunities associated with that social role (Moon 2020). Moreover, some historians have claimed that, in some eras, parents happily encouraged their boys to start lucrative sexual relationships with adult men. In Renaissance Italy, for example, the Franciscan Saint Bernardino of Siena scolded parents for acting as pimps to their own sons (Rocke 1988).

The publication of Wilson's *Sociobiology* in 1975 caused a substantial row that reverberated far beyond academia. Wilson primarily intended it to be a coffee-table science book about the evolutionary origins and mechanics of various social behaviors, but his ambitions surpassed a revolution in biology. He also wanted to offer a new foundation for the social sciences and, ultimately, to change the world by providing the knowledge needed to make better moral and political decisions. Some of his colleagues, however, were quick to read and dismiss the book as a bourgeois ideological treatise, charging it with, among many other things, racism, biological determinism, and ethnocentrism (Rose, Lewontin, and Kamin 1984).

Wilson's hypothesis about the evolution of homosexuality was also found wanting. Indeed, Wilson did not explain how one's homosexual orientation would contribute to the reproductive success of one's relatives. In nonhuman animals, helpers at the nest do not engage in homosexual behavior. In a later book, Trivers (1985) himself added that one would expect such helpers or sugar daddies to be asexual rather than homosexual, much like most sterile workers in eusocial animal species, like ants, bees, and naked mole rats. Wilson also offered no evidence that human male homosexuals did or do contribute to the reproductive success of their close relatives. More recent research has indicated that late twentieth-century Western homosexuals do not make such a contribution, or at least no more so than contemporary male heterosexuals (Bobrow and Bailey 2001; Vasey and VanderLaan 2012). Perhaps they make more money than male heterosexuals, but they prefer to spend it on themselves or on their partners, rather than on their siblings or their siblings' children (Hewitt 1995).

However, a newer series of scientific studies seems to provide some intriguing support for the sugar daddy hypothesis. Since the early 2000s, psychologists Doug VanderLaan and Paul Vasey have been conducting fieldwork among a particular population in Samoa, locally known as the fa'afafine (VanderLaan and Vasey 2014). Samoans recognize these people as belonging to (what Westerners would refer to as) a third gender—biological males raised to be feminine, who only have sex with individuals who identify as men. Results of this fieldwork indicate that the willingness to invest, as well as the actual investments, in the offspring of close relatives are higher in the fa'afafine than in both heterosexual men and women.[4] Interestingly, their investments also tend to be very selective, focusing mainly on young daughters of female siblings.

As VanderLaan and Vasey argue, this peculiar preference is largely predictable, at least from a kin selection point of view (VanderLaan and Vasey 2014, 1010–11; Vasey and VanderLaan 2010). First, the focus on daughters of female siblings, rather than on brothers and nephews, may be due to the problem of paternal uncertainty, as fathers, compared to mothers, have less certainty about their genetic kinship with their children. Second, the focus on young sisters and their daughters may be because these sisters have a larger residual reproductive potential than older sisters and their daughters. By definition, older sisters are closer to their menopause than younger sisters. In addition, younger children need more care than older children. Many studies on contemporary small-scale societies in rural areas indicate that mortality is highest among toddlers and decreases with age (Walker et al. 2006).

∴

All in all, there is no shortage of hypotheses about the evolution of human male homosexuality. In a fit of optimism, one might say that this abundance of hypotheses testifies to the imaginative power of evolutionary scientists. A cynic, by contrast, would view it as a symptom of an old sore, deriding the proclivity of some evolutionists to produce "just-so" stories—amusing and imaginative yet scientifically useless accounts, like Rudyard Kipling's burlesque children's tales about how the leopard got his spots or how the alphabet was made (Rose, Lewontin, and Kamin 1984, 261). Such criticism, however, is a bit too harsh. In the past two decades, some parts of some hypotheses about human male homosexuality have in fact been assessed through largely reliable and relevant data. The sugar daddy hypothesis has found some support in the Samoan fa'afafine, as has the sickle cell hypothesis in the enhanced fecundity in maternal relatives of male homosexuals.

Some evolutionary accounts of human male homosexuality may carry more weight than just-so stories, but that does not mean they are flawless. In the next two sections, we examine some of the dubious assumptions on which they are based, in particular, the homogeneity of human male homosexuality and the reproductive success of homosexuals.

HOMOSEXUALITY AND HOMOGENEITY

Some biological accounts of human male homosexuality are openly essentialist about their explanandum. In chapter 1, we defined the essentialist position as the belief that there is such a thing as the essence of homosexuality—a property or set of properties that all and only homosexuals have, have always had, and will always have. Some biologists indeed believe homosexuality to be a highly homogeneous category of people and allow for little or no variation in its genotype or phenotype through time and across cultures. Psychologists Qazi Rahman and Glenn Wilson (2003, 1341) provide an instructive example by claiming that "there is little cross-cultural and historical evidence that contradicts *the* [homosexual] *phenotype* observed in [contemporary] Western societies" (italics ours).

Similarly, some evolutionary thinking about homosexuality seems to imply that there is only one form of homosexuality, that is, modern homosexuality, or that that form is the only one worth investigating. Some studies focus exclusively on "male homosexual preference (MHP)," for example—a concept that they rather vaguely define as "sexual attraction to male partners even if female partners are available" (Barthes, Crochet, and Raymond 2015, 1). In the previous chapter, we distinguished between homosexual preference and orientation. We suggested reserving the term "preference" for transient or even single sexual choices and the term "orientation" for more continuous or permanent homosexual preferences. When evolutionists talk about "male homosexual preference," they are actually referring to an exclusive homosexual orientation. Such an orientation is indeed typical of modern homosexuality, and, by definition, it tends to exclude heterosexual reproductive activities. In this section, we argue that some evolutionary biologists are guilty of essentializing homosexuality in at least three different ways: by focusing more or less exclusively on (1) one of the many *forms* of homosexuality (i.e., modern homosexuality), (2) one of the many *causes* of homosexuality (i.e., genes), and (3) one of the many *dimensions* of homosexuality (i.e., orientation and identity).

What is wrong with an essentialist view of homosexuality? First, and despite concerted efforts, scientists have still not been able to identify any essential biological property or set of properties of homosexuality, and there is little reason to believe that they soon will. Established biological markers are few, for example, and reliable findings invariably come with the caveat that they do not explain the sexual orientation of all homosexuals, meaning we cannot consider them *essential* properties (Gavrilets, Friberg, and Rice 2018).

The fraternal birth order effect (FBOE), briefly mentioned in chapter 1, is a case in point. One of the few reliable findings in contemporary research on male homosexuality, the effect holds that the number of older brothers is a good predictor of the sexual orientation of their later-born brothers. In other words, the odds of a man being homosexual increase with the number of older brothers he has. Sexologist Ray Blanchard first described the FBOE in the early 1990s, showing that the effect reliably occurred with older brothers but not with older sisters, younger brothers, or younger sisters.

Some years later, his colleague Anthony Bogaert (2006) added that male homosexuality only correlates with the number of older biological brothers, not stepbrothers or adoptive brothers. He therefore suggested searching for the underlying mechanism in the mother's uterine environment, rather than in the family dynamics. More particularly, Blanchard and Bogaert hypothesized that some forms or instances of male homosexuality are due to the progressive immunization of some mothers to the antigens produced by their successive sons during the first months of pregnancy. This immunoreaction then disturbs the sexual differentiation of the fetal brain, causing a deviation from the male-typical pattern of neurodevelopment. One result of this deviation would be that later-born sons sexually prefer men to women. In the past decade, the FBOE has also proved to be cross-culturally reliable. Its occurrence has been established in populations as diverse as the Samoan fa'afafine (VanderLaan and Vasey 2011), Kinsey's mid-twentieth-century American interviewees (Blanchard and Bogaert 1996), and contemporary Iranian "homosexual" transgender women (Blanchard 2018, 4).

Blanchard has always been very open about the explanatory strengths and limitations of the FBOE. In a recent meta-analysis, he repeats that "only about 15–29% of the men in any given homosexual group can attribute their sexual orientation to the FBOE" (Blanchard 2018, 9). Many male homosexuals are firstborn or only children, or only have older sisters and no older brothers, and their sexual orientation is thus not explained by their fraternal birth order or the putative underlying mechanism. The conclusion that one of the most robust findings in contemporary research on human male homosexuality has limited explanatory power should make us cautious in expecting scientists to find any essential property soon.

In addition, it seems odd to expect, as essentialists do, that homosexuality would be a highly homogeneous category of people, when it seems likely that its biological etiology will be rather diverse. In chapter 1, we discussed the ongoing hunt for gay genes in ever-more ambitious genome-wide linkage and association studies. One recent study tentatively identified a number of DNA regions associated with male homosexuality—regions involved in neuronal function, neurodevelopment, and thyroid function (Sanders et al. 2017, 1). Blanchard and his FBOE collaborators state they gladly accept such evidence, but they also believe that some neurodevelopmental pathways involved in homosexuality are triggered by other etiologic factors, such as immunoreactions. Importantly, gay genes do not play any role in the maternal immunity hypothesis, at least in its current form (Blanchard 2018).

If some life scientists consider gay genes and maternal immunoreactions to be (partially) separate etiological factors, it is puzzling why they still expect the results to be identical. Of course, in the same way that all roads lead to Rome, different makeups might translate into a single phenotype. Moreover, contemporary philosophy of mind teaches us that mental states are multiply realizable, meaning that various physical states and properties can realize the same mental state or property. As Cordelia Fine (2010, 143) puts it: "The brain can get to the same outcome in more than one way." Similarly, essentialists could maintain that variation in the causes of homosexuality can still produce a rather homogeneous homosexual phenotype. Obviously, the success of this reply would crucially depend on whether there are no important differences in the phenotypes or the mental states that these hypotheses are focusing on. This brings us to our second objection.

A second problem with the essentialist view of homosexuality is that it fails to appreciate the diversity in human male homosexuality. It is true that in recent years awareness of this diversity has increased in biological circles, at least when it comes to diversity within the subgroup of modern homosexuals. Blanchard's meta-analysis provides an interesting example, as it reveals, in contrast to previous FBOE studies, that the magnitude of the effect is greater for the specific subset of "feminine" homosexuals: "Feminine homosexual males have more older brothers than non-feminine homosexual males, and non-feminine homosexual males, in turn, have more older brothers than heterosexual males" (Blanchard 2018, 10). Blanchard gracefully acknowledges that gender differences are "*one* of the most salient behavioral variables within the male homosexual population" (10; italics ours), so he seems to agree that they are not the only variable. Other recent work focuses on the endocrinological factors underlying the difference between "active" (or penetrative) and "passive" (or receptive) male homosexuals—a distinction already made by the Aristotelian author of the *Problemata*. One researcher recently suggested that these "subgroups of gay men may owe their sexual orientation to distinct bio-developmental mechanisms" (Swift-Gallant 2019, 26; see also chapter 2).

Still, some of the evolutionary literature on male homosexuality seems to overlook this diversity, which is all the more problematic since it claims to account not just for contemporary homosexuality but also for the entire evolutionary history of homosexuality. Moreover, it seems fair to say that, generally, there is much less attention to the diversity of homosexuality in the life sciences as compared to the human sciences, such as sex history and cross-cultural anthropology. Let us shift our focus from the diversity within the population of contemporary Western homosexuals to the historical and cross-cultural diversity of male homosexuality, which the human sciences unblushingly pluralize into many homosexualities (Murray 2000). Two of these forms of homosexuality are relevant to the rest of this chapter. Historians and anthropologists commonly distinguish between modern homosexuality, the form of homosexuality that twenty-first-century Westerners are most familiar with, and another form (out of many), which occurred, and continues to occur, in many premodern and contemporary non-Western societies, often referred to by researchers with terms such as "sodomy," "transgenerational homosexuality," and "pederasty." The differences between these two forms relate to a number of behavioral, psychological, and sociological variables.

A first variable is masculinity and femininity, which was already discussed by Blanchard in the context of modern homosexuality. Some historians claim that the first modern homosexuals contrasted sharply with their predecessors because of their eye-catching femininity. A striking example of this first feature of modern homosexuality is one of the early homosexual subcultures in the Western world: London's eighteenth-century "mollies." Mollies were adult homosexual men who performed a female role in so-called molly houses—a sort of private tavern where they acquired a female name, wore women's clothes, and spoke in high voices. Some mollies even presented with fake pregnancies and organized ritualized deliveries of wooden baby dolls (Trumbach 1998). In their sexual activities, mollies always played a "passive" role. Anthropologists might compare the contemporary Samoan fa'afafine to the eighteenth-century mollies in their overall femininity.

The femininity of modern homosexuality contrasts with some earlier and non-Western forms of homosexuality practiced by nonfeminine men with the express purpose of enhancing their masculinity. The ancient Greeks, discussed at the end of chapter 1, provide an illustrative example of this template. In many of their city-states, adult men, known as *erastai*, initiated older boys and adolescents, the *eromenoi*, into all sorts of manly virtues. Theirs was an intense relationship that was both pedagogical and physical, as a myriad of illustrations on Greek vases reveal (see fig. 3.2). Greek society explicitly expected the erastai to marry and raise children, but these men spent most of their time with their eromenoi, as they believed that any contact with women would make them more effeminate. Anthropologists have documented similar examples of nonfeminine male homosexuality in contemporary non-Western societies. In *Guardians of the Flutes* (1981), anthropologist Gilbert Herdt describes the Sambia people's small-scale society in late twentieth-century Papua New Guinea, which requires young boys and adolescents to regularly ingest the semen of a senior

Figure 3.2: A bearded *erastes* (*left*) feels a young *eromenos*'s genitals. The scene was immortalized on an Attic amphora (about 540 BC). © State Collection of Antiquities and Glyptothek, Munich, SH 1468 WAF. Photograph by Renate Kühling.

group member in order to achieve manhood, starting on average by the age of 8.5 years. He adds: "By the age of 11–12, the period of the second-stage initiation that advances the boys to the next level of the male hierarchy, the boys have become aggressive fellators who actively pursue semen to masculinize their bodies" (Herdt and McClintock 2000, 596). According to Sambia belief, sexual contact with women, especially contact before reaching full adulthood, threatens to pollute the male ideal of the hunter-warrior. The initiation of young men into manly virtues is thus a recurring theme in premodern and non-Western forms of homosexuality.

In chapter 1, we noted that some social constructivists make much of this distinction between modern and other kinds of homosexuality. In his early work, historian David Halperin (2002, 13) claimed, with much bravado, that "there was no homosexuality, properly speaking, in classical Greece, the ancient Mediterranean world, or indeed in most pre-modern or non-Western societies." The problem with such a dichotomy is that it saddles us with a new kind of essentialism, as Halperin himself acknowledged in a later work (Halperin 2002). There are in fact two assumptions underlying his early claim. Halperin assumes not only that there were no effeminate homosexuals in classical Greece but also that there is no Greek-style homosexuality in the contemporary world. Both assumptions seem unsupported by the available evidence. First, other historians have revealed that at least some premodern societies did have a form of homosexuality that bears some resemblance to modern homosexuals. In classical antiquity, for example, there were the *kinaidoi*—cross-dressing, effeminate homosexuals who were recognized as a separate type or identity in Aristotelian (and again in early modern Renaissance) physiognomy (Murray 2000, 155; Borris and Rousseau 2008). Second, historians, journalists, and anthropologists continue to reveal the enduring existence of "premodern" kinds of homosexuality in the contemporary world. The Sambia's practice can be matched, for example, with one that is locally known in today's Afghanistan as *bacha bazi*, which involves a sexual relationship between adolescent boys and adult men. The practice reached Western media when it was discovered that the US military had turned a blind eye to it during its invasion of the country (Chopra 2017). Similar "sodomitical" subcultures have been documented in early twentieth century New York (Chauncey 1997).

Status differences are a second variable that distinguishes modern and other kinds of homosexuality. Unlike many other kinds, modern homosexuality is distinctly egalitarian. Modern homosexuals tend to date men of about the same age and status, while many premodern societies explicitly frowned on such relationships.[5] We know from various sources, including then popular stage plays, that the ancient Greeks considered the *kinaidoi* laughingstock, presumably because they went against their culture of masculinity. Conversely, contemporary Westerners denounce and prosecute anyone who practices forms of homosexuality like Greek pederasty, Sambia sexual initiation rites, or Afghan *bacha bazi*. Still, this kind of homosexuality was very common in many parts of the world and throughout history, including medieval Egypt's Mamluks, early modern Japan's Samurai, and Renaissance Italy. About the latter, the American historian Randolph Trumbach (1998, 5) notes: "By the age of thirty, one of every two Florentine youths had been implicated in sodomy.... Sodomy was therefore so widespread as to be universal. But it was always structured by age." Religious authorities disapproved of the custom, but public opinion saw nothing wrong with it, provided the older lover played the "active" part. Among the Sambia, younger boys were until recently encouraged to fellate older adolescent boys, but men who continued to engage in homoerotic activity after the birth of their first child were (and maybe still are) considered deviant (Herdt 1999; Herdt 2019).

In Western Europe, this asymmetrical kind of homosexuality started losing its desirability and its visibility in the early eighteenth century, gradually giving way to the modern, more egalitarian kind of homosexuality, starting with the mollies in England and then spreading in all directions across the globe. Some historians claim that this transition is still ongoing—a hypothesis that finds some support, for example, in the persistence of phenomena like *bacha bazi*.

A third and final variable between modern and other kinds of homosexuality is homosexuals' attitudes toward their sexual activities. For the ancient Greeks and for cultures with similar practices, homosexuality was merely a set of behaviors that were embedded and encouraged in a complex world of manly virtues. For modern homosexuals, by contrast, sexuality is a key characteristic of one's identity. Historians variously refer to this process as the psychologization, subjectification, or interiorization of sexuality (Oosterhuis 2000). (See our discussion of the concept of a homosexual identity in chapter 2, which contrasts it with other dimensions of homosexuality such as behaviors, desires, preferences, and orientations.) Some historians claim to find evidence for this transition in historical court documents and criminal laws. In premodern times, the criminal prosecution of homosexuality was often based solely on the question whether there had been actual sexual contact between the defendants, particularly in the form of anal penetration ("res in re et effusio seminis"; Gerard and Hekma 1989). The mental states of the defendants, including their feelings, intentions, and desires, were deemed largely irrelevant (but see Hofman 2021). Conversely, modern law systems more often incorporate such states in sentencing homosexual partners (Van der Meer 1989).

Other historians disagree with this dichotomy, warning against a new kind of essentialism in which historical continuities are all too easily dismissed. In his monumental world history of homosexuality, Louis Crompton (2003, xiv) notes:

> Even the idea of a sexual identity is not uniquely modern. Aristophanes expressed it plainly enough in the *Symposium*, and the Romans used it, in a limited sense, in their concept of the *cinaedus* ("faggot") [the Roman form of the Greek *kinaidos*], who was certainly a distinct sort of person.... Even in medieval times, when the view of same-sex relations as sins and crimes predominated, a French poet could make his heroine speak of "men of that sort" (*de ce métier*), that is, of a certain kind of individual.

It is of course notoriously difficult to obtain a truthful picture of homosexual lives in any era of the past. The predominance of particular source materials can create an illusion of homogeneity, further fueled by the fact that most societies had or have a distinct preference for one kind of homosexuality and an equally distinct loathing for others. If the latter forms are rarely referenced to, it may be because of their relative rarity or simply because one could not speak their name.

∴

All things considered, it is fair to say that the population of human male homosexuals at any one point in time is not as homogeneous as some biologists and evolutionary scientists would have us believe—a conclusion that, according to one commentator, is "obvious . . . but rarely admitted" (Jannini et al. 2010, 3247). Acknowledging this heterogeneity, however, seems vital for the biological gay sciences, since it likely constitutes one of the main reasons why it is so difficult to produce reliable data. Moreover, it may have important consequences for evolutionary theorizing of human male homosexuality, as different forms of homosexuality will probably require different evolutionary explanations. In the final sections of this chapter, we will illustrate this point by detailing an evolutionary account of premodern homosexuality and an alternative evolutionary account of modern homosexuality. We first, however, explore a second erroneous assumption built into many evolutionary hypotheses of human male homosexuality.

HOMOSEXUALITY AND REPRODUCTIVE SUCCESS

One of the central assumptions of many evolutionary accounts of human male homosexuality is that homosexual men have far fewer offspring than heterosexual men. Alongside claims about the genetics of homosexuality and the historical invariance of its phenotype, this assumption is one of the buttresses of the paradox we presented at the beginning of this chapter. If it is true that human male homosexuality has always existed, if genes contribute to its development, and if homosexuals leave little or no offspring, then why does it still exist?

Evolutionary scientists routinely describe human male[6] homosexuality as "a reproductively costly trait" (Barthes, Crochet, and Raymond 2015, 1) or even as "the antithesis of reproductive success" (Potts and Short 1999, 74). Some are more cautious, speaking of an "*apparently* detrimental trait" (Camperio-Ciani, Battaglia, and Zanzotto 2015, 2; italics ours) or claiming that homosexuals "are *presumed* to have lower fitness and reproductive success" (Barron and Hare 2019, 1; italics ours). Still others acknowledge that "it is likely that in some cultural contexts, nonheterosexuality decreases reproduction to a lesser extent" (J. Bailey et al. 2016, 77), leaving open the possibility that some "homosexual men [do] have some direct offspring" (Nila et al. 2018, 2456). All in all, contemporary researchers still seem to echo the words of George Evelyn Hutchinson, who briefly reviewed some rather anecdotal fertility evidence and concluded that "such meager evidence supports the commonsense view that homosexuality in western cultures is likely to be correlated with a low fertility" (Hutchinson 1959, 82).

For twenty-first-century Westerners who are most familiar with the modern kind of homosexuality, the assumption does appear to be a matter of common sense. If one defines homosexuality as an exclusive sexual relationship with an individual of the same sex, then most often homosexuals will not produce offspring.[7] "You can't inseminate another male," the gay psychologist Jesse Bering (2014) dryly notes. "Believe me, I've tried." This accepted logic is perhaps one of the main reasons why there has been so little research on the actual reproductive fitness of homosexual men, both past and present.

The few available studies do in fact suggest that contemporary homosexual men have fewer children than contemporary heterosexual men. An Australian study from the early 1970s found that among a group of ninety-five homosexual men above age forty, they had thirty-seven children in total, which is significantly lower than the number of children expected for a group of heterosexual men of the same age (Moran 1972). (In 1972, the fertility rate was around replacement level.) Twenty-five years later, researchers reported comparable numbers in Australia, at least for older homosexual men (defined here as men above age forty-nine who self-identify as homosexual) (Van de Ven et al. 1997). Nearly two-thirds of them were or had been married, while just over half of them had children. Younger homosexual men were even less likely to have children, which may be evidence of a change in reproductive success among Australian homosexuals at the turn of the century.

One finds hints of a similar change in contemporary American studies on homosexuality and reproduction. In 1978, Alan Bell and Martin Weinberg, two researchers at the Kinsey Institute, estimated that the reproductive success of homosexual men was one-fifth of that of heterosexual men. Less than twenty years later, a new study found even lower numbers for the reproductive success of homosexual men, approximately one-tenth of that of heterosexual men (Hamer and Copeland 1994). That same year, a representative general survey of gays and lesbians found that one-quarter of gay men were fathers, compared with more than half of straight men (Yankelovich Partners 1994). In 2005, an English study reported an all-time low in fertility numbers, with gay men producing only 0.002 offspring on average, compared with 0.36 offspring produced by straight men (King et al. 2005). In a more recent study on firstborn men, however, the numbers were much higher: 0.16 offspring for (firstborn) gay men and 0.53 for (firstborn) straight men (Schwartz et al. 2010). Unfortunately, data about the reproductive success of non-Western contemporary homosexuals (or similar populations) are even more rare. One study on the fa'afafine noted that they had no children at all (Vasey, Parker, and VanderLaan 2014).

Overall, the few available studies on homosexuality and reproduction seem to be consistent with evolutionary theorizing in indicating that, compared to heterosexual men, homosexual men have fewer children, if any. However, there are good reasons to believe that fertility numbers from contemporary Western (and even some non-Western) homosexuals are not representative of all homosexuals at all times and in all places, as some evolutionists assume or argue (e.g., Camperio-Ciani, Battaglia, and Zanzotto 2015, 4; VanderLaan, Ren, and Vasey 2013). It is curious to note that, in the context of the debate about sickle cell hypotheses, some biologists questioned the representativeness of data on the reproductive output of maternal relatives of contemporary male homosexuals but not the representativeness of data on homosexuals themselves: "Would contemporary mothers of firstborn homosexual sons have been relatively fecund 100, 1,000, or 10,000 years ago? It is obviously impossible to know this, but our ignorance of the mechanism for their present increase requires caution. We certainly cannot assume that it would have operated with the same reproductive consequences that it does today" (Rieger et al. 2012, 530).

Would firstborn or later-born homosexuals have been relatively fecund one hundred, one thousand, or ten thousand years ago? Changes in reproductive behavior in the West since the early nineteenth century make it difficult to use Western samples to determine the historical fitness effects of human traits and behaviors. In the West, for example, there seems to be a negative correlation between socioeconomic status and reproductive success (Skirbekk 2008, but see Hopcroft 2006 for some nuance), while the correlation is positive in many other parts of the world. Likewise, many researchers believe that the reproductive success of male homosexuals differs across time and cultures, if only because Western societies are outliers with regard to social norms about reproduction and with regard to the impact of these norms on the reproductive behavior and choices of homosexuals. In addition, one could argue that the apparent decline in fertility numbers of Western homosexuals in the past half century is due not only to a general decline of the Western fertility rate but also to the increasing tolerance toward the modern kind of homosexuality, at least in some parts of the Western world. We will further discuss this issue in the last section of this chapter.

More importantly, there is some positive evidence to suggest that what little we know about the fertility of male homosexuals in the past does contradict our commonsense assumptions and the data used to support them. One of the more surprising findings of recent historical and anthropological research on homosexuality is that many premodern and non-Western homosexuals had (and continue to have) sexual relationships with both women and men, mostly boys or adolescents. Many of them were indeed married and had children (Murray 2000; Berman 2003). In ancient Greece, for example, Spartan boys or eromenoi were drilled under the eye of their older lovers, the erastai, to become good warriors. The erastai themselves were married men. Eventually, most of the boys would also marry and were expected to have intercourse with their wives every now and then. The remaining nights they spent with their erastai, who often acted as the newlyweds' Maecenas for some time after the marriage, or with their own eromenoi. Indeed, only the eromenoi who married and raised children were allowed to become erastai themselves: "Exclusive pederasty was negatively sanctioned, but pederasty was expected" (Murray 2000, 40).

In short, male homosexuality frequently involved, and still does involve, married men. While the idea of a homosexual man marrying a woman may seem strange to some of us today, not only did such marriages frequently occur in many ancient societies, but they continue to occur in many contemporary societies, including the United States (Bagley and Tremblay 1998). In Japan, for example, marriageable women often read gay magazines because they contain personal ads from homosexuals whose families and employers are urging them to marry and beget children: "So long as those obligations [marriage and parenthood] are met, one's sexual activity is not anyone else's legitimate concern" (Murray 2000, 398). One finds a similar example in the fa'afafine. Recent studies reveal remarkably low fertility figures in this population of Samoa, but it is important to note that fa'afafine prefer to have sex with "gynophilic," or heterosexual, men. To our knowledge, however, researchers have not studied if, and to what extent, the latter's sexual activities affect their reproductive success.

Inevitably, then, the debate about homosexuality and fertility brings us back to the debate about homosexuality and homogeneity. If it is true, as we previously argued, that there are often strikingly diverse forms of homosexuality, then it is possible that they also differ with regard to fertility. In this context, it is interesting to note that studies on contemporary homosexuality that include bisexual men who have sex with both men and women yield much higher fertility numbers. In the early 1990s, for example, a Japanese study revealed that of 655 homosexual and bisexual men, 83 percent were fathers (Isomura and Mizogami 1992). The comparable Australian study mentioned above did not include bisexual men and found that only 56 percent were fathers (Van de Ven et al. 1997).

These and other observations suggest that for many homosexual men, both past and present, homosexual activities are complementary to, and not a replacement of, reproductive sexuality. More generally, it seems rash to assume that, on average, premodern and non-Western homosexuals had as little offspring as contemporary modern homosexuals. To a certain extent, this conclusion subverts the power of our paradox. As biologist Paul Ehrlich (2000, 198) states: "Many [gay men] do marry and have children, and many may be very sexually active with both sexes. Selection against the allele . . . predisposing to homosexual behavior might thus be very weak." (See also Kirkpatrick 2000.) However, even some historians and anthropologists admit that various kinds of exclusive male homosexuality can be found in some premodern and non-Western societies, thus leaving open the question how it is that exclusive homosexuality has managed to elude natural selection.

In our view, the real problem is not so much that some evolutionary scientists consider human male homosexuality as a Darwinian paradox but rather that they overestimate the importance of that paradox. In doing so, they focus too exclusively on only one of the many *forms* of homosexuality (i.e., modern homosexuality), one of the many *causes* of homosexuality (i.e., genes), and one of the many *dimensions* of homosexuality (i.e., orientation and identity) (for similar criticism, see Savolainen and Hodgson 2016). In the last two sections of this chapter, we shift our attention to the evolutionary specifics of other forms, causes, and dimensions of homosexuality.

THE ALLIANCE FORMATION HYPOTHESIS

What is the function of human male homosexuality? It seems unlikely that modern homosexuality is adaptive, even if it may be related to adaptive advantages in the close relatives of homosexual men, as in the sugar daddy hypothesis and the various sickle cell hypotheses. Once we realize, however, that unlike modern homosexuality, some homosexual behaviors, desires, and even preferences and orientations are perfectly compatible with reproductive behavior, we are then confronted with the question what possible adaptive advantages these features have for homosexual men themselves. Why is it that some men have, desire, or prefer sex with men, even if they have access to women, are married, and have children? The answers to this question are probably as diverse as the topic they touch on, especially when it comes to nonhuman male homosexuality (Vasey and Sommer 2006; N. Bailey and Zuk 2009; Adriaens and De Block 2016). One answer in particular goes some way to account for the diversity of human male homosexuality: the *alliance formation hypothesis* (Kirkpatrick 2000; Muscarella 2000; Vasey and Sommer 2006; De Block and Adriaens 2004; Adriaens and De Block 2006).

The alliance formation hypothesis of male homosexuality claims that homosexual behaviors and other states were selected for in many animal species, including humans, because they establish, maintain, and strengthen strategic alliances among nonkin. These alliances enhance the individual's chances of survival and reproduction relative to rivals who are less inclined to engage in such sexual behavior. The alliance formation hypothesis has been shown to hold true for a variety of nonhuman animals, including bottlenose dolphins (*Tursiops truncatus*; Mann 2006), graylag geese (*Anser anser*; Kotrschal, Hemetsberger, and Weiß 2006), Sumatran orangutans (*Pongo pygmaeus abelii*; Fox 2001), and olive baboons (*Papio anubis*; Smuts and Watanabe 1990). Studies on social primates have shown that low-status males often negotiate their position in the group hierarchy by means of homosexual activities, then cease such activities as soon as they achieve higher status. In langur monkeys (*Presbytis entellus*), for example, individuals belonging to so-called bachelor bands engage in homosexual behavior before conducting a collective raid on the dominant male's harem. According to one study, 95 percent of all sexual activities in a langur bachelor band are homosexual activities, while dominant males have sex only with females (Sommer, Schauer, and Kyriazis 2006).

Human men may likewise use homosexual behavior to form bonds and alliances in situations where cooperation is vital. Many examples from the historical and ethnographical literature on human homosexuality seem to confirm the basic predictions of the alliance formation hypothesis (although we acknowledge that these examples might have been, to some extent, cherry-picked).

Homosexuality is known to have been common in unusual circumstances where bonding was vital, such as wars and long expeditions. In populations as diverse as the Thebans of ancient Greece, the Mamluks of medieval Egypt, the Samurai of preindustrial Japan, and the Sambia of today's Papua New Guinea, sexual activities and mental states often served to solidify long-lasting companionships (Herdt 1994; Murray 2000). In Plato's *Symposium* (1925, 178e), for instance, Apollodorus famously claims that "a state or an army should be made up of lovers and their loves"—a notion that likely inspired the formation of the legendary Sacred Band of Thebes, an elite troop of 150 homosexual couples within the Theban army (Crompton 2003). A similar kind of wartime homosexual camaraderie also characterized Japanese Samurai culture. Originally, the Samurai formed a separate caste of warriors, and every warrior, although married with children, was supposed to initiate a young boy page into the manly virtues of the Samurai—virtues that included loyalty, determination, and honor. A Samurai also provided his page with emotional and, if needed, financial support. As one commentator notes, "As in marriage, sex was only one element of the man-boy relationship" (quoted in Murray 2000, 80). Although sex was but "one element," it was one with a huge impact: Samurai warriors, like their counterparts in Thebes, preferred the company of their sex partners while fighting their enemies. For them, only a sexual bond between soldiers could vouch for the extreme loyalty needed in battle (see fig. 3.3).

Alliances may be vital in wartime, but they are certainly useful in daily life too. A number of anthropologists and historians have noticed that homosexuality can bring about a form of solidarity that isn't based on kinship. The Gebusi people in Papua New Guinea provide an excellent example. In the early 1980s, the American anthropologist Bruce Knauft found that many Gebusi men often engaged in informal homosexual activities: "All men in the settlement have a strong sense of relaxed friendship and communal camaraderie. Their network of past or present homosexual relations reflects this spirit of diffuse affections, encompassing men who cannot trace specific kinship ties to one another" (Knauft 1985, 32).

Another example of the connection among homosexuality, cooperation, and reciprocity can be found in the "Greek-style" homosexual relationships between adult men and adolescents in fifteenth-century Florence. Although such relationships were socially asymmetrical, they often had important benefits for both parties. Parents encouraged and dressed up their sons to attract older men, so the latter would shower them with gifts and money; while some adolescents were in it for the money, the adult men also needed them to function as status symbols. Indeed, the chances of getting promoted to important civic offices crucially depended on the quality of one's "boy(s)" (Rocke 1996). Again, homosexual alliances demonstrably improved the living conditions (and presumably the survival) of both partners.

A recent experimental study provides further and more systematic support for the alliance formation hypothesis (Fleischman, Fessler, and Cholakians 2014). In the study, experimenters subtly alerted heterosexual male test subjects to the importance of having male friends and allies. As a result, the test subjects displayed much more openness to sexually experiment with other men. The effect was most noticeable in men with high levels of progesterone—a sex hormone known to be involved in the formation of social bonding.

While homosexuality can play a vital role in establishing alliances, there are still plenty of alliances that are not built on homosexual activities. Unlike

Figure 3.3: In Samurai culture, the love between an adult and his page was known as *nanshoku*—a kind of homosexuality often depicted in erotic illustrations (*shunga*). In this drawing, the sixteenth-century feudal lord Hideyoshi kneels for a page while stroking his wrist (1804). © Trustees of the British Museum. The British Museum Kitagawa Utamaro, Mashiba Hisayoshi, Kyowa Era, Bunka Era, 1942, 0124,0.13.

many other animals, humans have various powerful nonsexual means for bonding, such as shared language and rituals (Muscarella 2000). Nevertheless, sexuality remains one of the most effective symbols of loyalty and affiliation (Kirkpatrick 2000). In the previous chapter, we gave the example of diddling—a form of genital touching among adult male olive baboons (*Papio anubis*) often seen in the context of greeting. One interpretation of this behavior is that it is a powerful symbol of trust and solidarity. The willingness to be touched intimately can be considered an honest signal, since it involves a physical interaction that is much more vulnerable and therefore much more reliable (Zahavi 1977) than, for example, verbal interaction. Talk is often cheap, and actions speak louder than words.

The alliance formation hypothesis has a number of interesting advantages. First, it shifts the evolutionists' attention from the modern, exclusive kind of homosexuality to other kinds of homosexuality that many evolutionary scientists have tended to ignore. Second, in doing so, it opens up the possibility of a joint venture between the life sciences and the human sciences, as the kind of homosexuality explained by the alliance formation hypothesis is one that historians and anthropologists are more familiar with. Such a joint venture would be unique in a scientific world still riddled with antagonism between evolutionary biologists and human scientists. We mentioned one example of this antagonism in chapter 1, when Rahman and Wilson (2003) easily dismissed social constructionist thinking about human male homosexuality. Third, the alliance formation hypothesis is a comprehensive hypothesis, in that it accounts for a portion of homosexual behaviors, desires, preferences, and orientations in both humans and other animals. This comprehensiveness is rare in the literature, as many evolutionary hypotheses about animal homosexuality are inapplicable to humans (Adriaens and De Block 2016), and vice versa, as in the case of the sickle cell hypotheses and the sugar daddy hypothesis.

Finally, the alliance formation hypothesis can help us make interesting and testable predictions[8] about the increasing occurrence of exclusive kinds of homosexuality in rapidly urbanizing areas, such as nineteenth-century Europe. In chapter 1, we mentioned that some social constructivists believe that modern homosexuality is an entirely new social phenomenon produced by nineteenth-century sexual categories and classifications, or at least that its visibility and popularity (and desirability) increased rather rapidly at the beginning of the early nineteenth century. How can this apparent increase be explained? European cities have witnessed homosexual subcultures since as early as the sixteenth century (Perry 1989). For those who spend most of their time in densely populated areas, rather than in traditional, kin-based communities, creating and confirming new alliances with strangers may be of special importance. The alliance formation hypothesis predicts that urban anonymity and concern with nonfamilial pursuits, much like wartime conditions, would foster homosexual companionships.

One of the limitations of the alliance formation hypothesis is that it focuses heavily on sexual behavior while ignoring the construction of cultural norms regarding sexual behavior. When it comes to behaviors, humans are more flexible than any other animal species because they largely rely on complex social learning mechanisms—mechanisms that, like genes, have been preserved by natural selection (Richerson and Boyd 2005). The alliance formation hypothesis predicts that any environment necessitating the use of alliances will increase the occurrence of homosexual behavior, regardless of what people believe about such behavior. However, a society's beliefs and norms about homosexuality do play an important role in understanding homosexual behavior in that society. Biologist R. Craig Kirkpatrick (2000, 397) admits as much: "Humans are quite plastic in conforming to social institutions. In some societies of Melanesia, in 17th-century Japan, and in classical Athens, men have been expected to find men sexually attractive, and on the whole they have done so." Social scientists, including social constructivists, emphasize how knowledge and concepts constitute people, and it is unclear how the alliance formation hypothesis contributes to our understanding of how this occurs. There are other evolutionary approaches, however, that shed a light on this issue, one of which we discuss below.

THE KIN INFLUENCE HYPOTHESIS

Evolutionary theory is a house of many mansions, as we hope this chapter has already amply illustrated. In this section, we focus on one particular evolutionary approach to human affairs: dual inheritance theory (DIT). This style of evolutionary analysis emphasizes the effect of social learning and social influence on human behavior. DIT advocates believe that the variation and variability in human behavior is best understood as the product of two interacting inheritance systems—genetic inheritance and cultural inheritance. Like genes, cultural information evolves as it is transmitted from generation to generation, or from one mind to another, and much like the process of genetic copying, this transmission gives rise to endless new variations. Some of these variations are more useful than others, and through natural selection, they soon spread all over a population. Many of the products of culture are thus adaptive, and dual inheritance theorists argue that our capacity for culture is in fact a biological adaptation. Here we ask whether DIT can help us understand and explain the origin and rise of modern human male homosexuality.

Evolutionary scientists have always quarreled with one another over the relative importance of, and the interaction between, social and cultural forces and biological forces. The work of Edward O. Wilson represents one extreme end of this debate. He focused rather exclusively on the biological determinants of various features in the social and cultural life of humans and other animals. In *On Human Nature*, he is quite straightforward about this focus: "The genes hold culture on a leash. The leash is very long, but inevitably values will be constrained in accordance with their effects on the human gene pool" (Wilson 1978, 167). Less than a decade later, anthropologist Robert Boyd and biologist Peter Richerson teamed up to publish their counterargument, *Culture and the Evolutionary Process* (Boyd and Richerson 1985). They provided a well-wrought, math-based analysis of all sorts of interaction processes between genes and culture—processes that lead to the current diversity within and among human populations.

Boyd and Richerson agree with Wilson that certain cultural phenomena are the product of our biological evolution, but they also emphasize the reverse: sometimes our biological evolution is directed by cultural developments. A classic example is the evolution of lactose tolerance. Careful statistical work has indicated that a history of dairying is the best predictor for the prevalence of lactose tolerance, which genetics has revealed is controlled by a single dominant gene. Approximately seven thousand years ago, lactose tolerance began to spread over European and some other populations—an evolution more or less paralleled by the domestication of cattle, including dairy stock. The common consumption (and indeed the abundant presence) of milk and dairy products created an enormous, culturally caused selection pressure on the genetic material involved in the production of lactase, the enzyme used in breaking down lactose.

Dual inheritance theorists also argue that the evolution of culture can be analyzed using methods similar to those used by population geneticists to analyze genetic change in a population. For example, it is possible to identify evolutionary forces that influence the speed and direction of cultural change. Natural selection is not only important for genetic evolution but also a force in cultural evolution. People who have better information about how to get food have better survival chances and greater reproductive success. Of course, other forces operate on cultural evolution as well. Individuals often choose, consciously or unconsciously, which cultural variants to adopt. The rules that people use, and that bias their decisions, act as forces that affect the frequency of cultural variants. Examples of such biases include the conformist bias ("do what most people do") and the prestige bias ("do what the most successful people do"). One of DIT's basic ideas is that these transmission biases in cultural evolution are themselves the product of genetic evolution (see Heyes 2016, however, for a largely culturalist explanation of these transmission biases). In a nutshell, the genes involved in cultural transmission biases that regulate the spread of cultural information were naturally selected, and this information, in turn, can influence the evolution of genetic variants. According to DIT, genes and culture do indeed evolve in a complex and interactive way.

DIT argues that cultural evolution is what has made our species so successful. Without culture, we would not have been able to thrive in such diverse habitats as the Amazon rain forest and Alaskan tundra (Richerson and Boyd 2005; Henrich 2017). Wilson may have rightfully said that genes hold culture on a leash, but he dramatically underestimated the length of the leash. He believed that genetics restrains culture in order to reduce the risk of learning maladaptive behaviors due to carelessness or manipulation on the part of competing conspecifics. Such adaptations might cause an organism to experience pleasure when considering an adaptive choice and disgust when considering a maladaptive choice. It has long been known, for example, that newborn infants prefer sweet tastes to bitter and sour tastes, which suggests that humans have genetically evolved taste preferences (Nisbett and Gurwitz 1970). However, the fact that many infants grow up to prefer double espressos and lemon tea without sugar shows how easy it is to override these preferences. Dual inheritance theorists emphasize that people draw little benefit from genetically evolved food preferences because they can simply learn from others what is good to eat. Evolved preferences could even reduce survival by causing individuals to avoid safe and nutritious food that happens to taste bitter or sour.

Fellow humans with whom we grow up and live influence not only our taste preferences and food customs but also our beliefs and interests, including our beliefs about sexual attractiveness. For example, surveys of Zulu communities in South Africa show that men prefer a female shape that reflects a much higher body mass index than that preferred by Western men. However, a sample of Zulu men who had immigrated to London were found to have developed a taste for thinner women within a few months of arrival (Tovee et al. 2006). This suggests that humans may lack strong genetically evolved sexual preferences because, for most of human evolution, individual preferences played a limited role in their procreative choices.[9] Studies of mating behavior in the kind of small-scale societies in which most humans lived until about two hundred years ago reveal that reproduction largely took place within marriage, and many areas were so sparsely settled that most people did not have many options for marriage partners. Meetings between possible partners were organized by networks of kin, and friends and parents exercised considerable influence over whom their children, especially their daughters, would marry (Apostolou 2010). Some evolutionary anthropologists have argued that the widespread occurrence of arranging marriages among hunter-gatherers means that the practice likely originated before anatomically modern humans left Africa more than fifty thousand years ago.

What, if anything, can DIT tell us about homosexuality? In the past two decades, psychologist and dual inheritance theorist Lesley Newson has attempted to answer this question (Newson et al. 2005; Newson and Richerson 2009; Adriaens, De Block, and Newson 2012). Her kin influence hypothesis argues that changes in reproductive norms during a process known as economic development or economic modernization brought to the fore a new form of homosexuality—modern homosexuality. In brief, her evolutionary hypothesis about homosexuality agrees with the idea of human scientists that social learning has had, and continues to have, a big impact on the development of human sexuality.

According to Newson, modern homosexuality relates to economic modernization—a catchall term referring to the complex aggregate of monetary, technological, institutional, cultural, and social changes that first occurred in industrializing late eighteenth- and early nineteenth-century England and has continued to colonize the rest of the world ever since. This process had and still has an enormous impact on the structure of human networks and communities, as well as on the way they learn from and influence one another.

Evolutionary scientists have a special interest in societies that have not yet gone through this process of economic modernization. While these societies are (culturally) highly diverse, they have one thing in common: members' reproductive decisions suggest they are competing to maximize their fitness, just as evolutionary theory predicts (Borgerhoff-Mulder 1988). Why is that? Prior to economic development, the extended family was the dominant social institution for most individuals (K. Davis 1937). The family provided education and employment, among many other things. Most people were illiterate, so social information was largely communicated through face-to-face interactions among people who lived in the same region (Watkins 1990). Membership in a family provided an important social identity, and family members all prioritized preserving the family. One would therefore expect these family-based communities to maintain norms that encourage or promote the welfare and reproductive success of the family as a whole (Anderson 1983).

Darwinists use the term "inclusive fitness" to refer to the reproductive success of such a family. Maximizing inclusive fitness does not equal producing children profligately. Social norms about who reproduces and when constrain the rate of reproduction to match the availability of both the physical and social resources that mothers must draw on for help in raising their offspring. Those who do not reproduce themselves are expected to help their relatives. (Here we can find inspiration for the sugar daddy hypothesis, as explained earlier in this chapter.) In this way, families produce approximately the maximum possible number of surviving offspring, given their circumstances.

European populations, which were the first to undergo economic modernization, have now experienced over two centuries of rapid and ongoing cultural change (Inglehart and Welzel 2005). What has this development brought about in Europe? One of the most striking changes is the gradual disintegration of family ties. Increasingly, people are educated, entertained, and employed by nonfamily members. They spend more time away from home and become increasingly connected to the wider world through books and other new media. They form new networks, with corresponding new identities and social interactions that, in their turn, bring new influences and new ways of influencing. There is no reason to expect these nonfamily groups to generate or maintain norms that promote the family or encourage reproductive success. Little by little, these norms will die out (Newson and Richerson 2009).

Statistics reveal the result: a marked decline in the fertility of economically modernized societies. Indeed, one of the first new cultural beliefs following economic modernization was that it is prudent to *limit* family size. The resulting changes in fertility were accompanied by changes in parenting norms and gender roles and, of course, the long sexual revolution in the second half of the twentieth century (Weeks 2007). Last but not least, economic development has triggered the construction of new beliefs about and attitudes toward homosexuality. In societies that have not yet modernized economically, such beliefs and attitudes are very diverse, ranging from stark repudiation to shy encouragement or even glorification (Kirkpatrick 2000). In economically modernized societies, however, the relationship between fertility and the acceptance of homosexuality is inversely proportional: as the average fertility of the population goes down, the acceptance of homosexuality goes up (Adriaens, De Block, and Newson 2012, 89). European countries, or those with populations of mostly European descent, first experienced fertility decline. In these countries, fertility is currently low, mostly below two children per woman. These countries also exhibit the greatest acceptance of homosexuality. Some countries, mostly in East Asia, where fertility began to decline in the middle of the twentieth century show moderate levels of acceptance. Most countries where fertility only recently began to decline have very low tolerance of homosexuality.

Newson believes it is no coincidence that modern homosexuality only started to flourish in economically developed countries. The cultural revolution, and particularly the complex of new reproductive norms and practices, proved to be an ideal breeding ground for this kind of homosexuality. The increasing acceptance of homosexuals allowed them to form a critical mass, and the lack of strict reproductive norms means that modern homosexual men are more or less respected in their decision not to marry a woman and have children. The resulting subculture has created a completely new identity. While some aspects of modern homosexuality may also be found in premodern societies, according to Newson, modern homosexuality itself could only develop after the traditional family lost its normative importance.

The kin influence hypothesis greatly reduces our concern about some of the evolutionary literature's lack of interest in the diversity of human male homosexuality. Interestingly, it also fits in with the social constructivist literature discussed in chapter 1. There are striking similarities, for example, between Newson's hypothesis and what we have called ontological constructivism, which holds that modern homosexuality is a new phenomenon brought about by a society's sexual norms, categories, and classifications. Both views are based on the assumption that human sexuality is much more malleable than some scientists are willing to assume. (Newson even considers such malleability adaptive.) They may differ on the nature of the social forces affecting human sexuality, but they both believe in the power of cultural norms and beliefs to create new phenomena, including a form of homosexuality that differs from all other forms in being exclusive, egalitarian, and identity based. Overall, the kin influence hypothesis offers a plausible evolutionary account for the emergence of a modern kind of homosexuality and for the relative absence of such homosexuality before the industrial revolution.

Still, we have a few objections to this hypothesis. The first relates to the role Newson assigns to economic modernization. It is indeed striking how the fertility decline follows industrialization. However, the institution of marriage and other sexual norms had been "modernizing" long before the industrial revolution. Joseph Henrich, one of the most influential DIT researchers, has argued that the roots of what he calls the "Western, Educated, Industrialized, Rich, and Democratic," or WEIRD, mind are to be found in the Christian church's restrictions on marriages and date back to the early Middle Ages (Henrich 2020). These restrictions dissolved or radically transformed many of the institutions based on kinship, which had been so adaptive in most sedentary societies. The church eliminated polygynous and cousin marriages that bound large kin-based groups together. The nuclear family remained key, but all other kin-based relations became much less important. Henrich convincingly argues that this paved the way for new kinds of associations among people, such as guilds, and for a more individualistic psychology. A cross-country comparison reveals, for instance, that "countries with a longer exposure to the medieval Western Church have lower rates of cousin marriage; countries with lower rates of cousin marriage have a more individualistic and impersonally prosocial psychology; and countries with a longer exposure to the medieval Western Church have a more individualistic and impersonally prosocial psychology" (Schulz et al. 2019, 707). Compared to the rest of the world, and compared to the pre-Christian West, our modern Western identities are defined much less by our place in a lineage or clan and much more by what we do, feel, and desire, as well as by our place in voluntary associations. According to Henrich, this change led to urbanization and industrialization, eventually culminating in material wealth and democratic institutions. In other words, Newson is likely correct that economic modernization created new sexual norms, but it was the medieval church's norms about incest and marriage that were ultimately responsible for the rise of individualism in the West and its resulting economic developments. Ironically, the church's obsession with incest may have eventually led to the sexual revolution of the twentieth century that, in its turn, did away with so many Christian sexual prescriptions and prohibitions, including those against homosexual desires and behaviors.

Our second objection concerns the relationship between cultural norms and homosexuality. How did these norms create a new form of homosexuality? Newson claims that changing norms lead to changing behaviors that, in their turn, lead to changing desires and preferences. However, she does not discuss the mechanisms behind all this. Through Henrich's account, and through our own classification of the dimensions of homosexuality in the previous chapter, we can complement Newson by saying that the cultural evolution of the Western mind may have created new kinds of sexual desire, preference, and orientation. Particularly important in this context is the emergence of a markedly individualistic psychology, with its focus on introspection and intentions (Henrich 2020, 50–52). So-called WEIRD people placed importance on thinking about one's personal identity, desires, and aspirations in life, and this emphasis thus revealed previously unimaginable possibilities. One such possibility is being gay, particularly in the sense of deriving one's identity from one's sexual behaviors and desires. The cultural evolution of the Western mind did not only reveal new sexual possibilities but also increased the recognition and visibility of these possibilities, which finally resulted in the new form of modern homosexuality. It is possible that the strong persecution of homosexuals contributed to these developments. As one author speculates, "The relatively high prevalence of exclusive homosexuals may reflect aspects of the present environment; for example, condemning the behavior may create a need to 'choose side'" (Grinde 2022, 455).

CONCLUSION

Many evolutionary scientists overestimate the importance of the form of human male homosexuality with which they, as twenty-first-century Westerners, are most familiar. In this book, we describe this form as modern homosexuality, characterizing it as an identity-based, egalitarian, exclusive, and (very often) feminine kind of homosexuality. However, historians and anthropologists alert us to the fact that modern homosexuality is not particularly representative of all kinds of homosexuality at all times and in all places. It seems, then, that much of the evolutionary literature too quickly passes over the historical and cross-cultural diversity of human male homosexuality. In this chapter, we tried to compensate for this one-sidedness by discussing two recent Darwinian hypotheses that go some way to account for this diversity.

It is our hope that such hypotheses will attract more attention in the future, if only because, implicitly or explicitly, they help spread a message that some philosophers believe to be at the heart of contemporary evolutionary theory, which the American philosopher David Hull (1998, 388) summarizes as follows: "If evolutionary theory has anything to teach us, it is that variability is at the core of our being. Because we are a biological species, and variability is essential to biological species, the traits which characterize us are likely to vary, our own essentialist compulsions notwithstanding." To be fair, Hull was alluding to the frequent disregard in the field for sexual orientations other than heterosexuality, by presenting homosexuality, for example, as some kind of defect or anomaly. In our view, however, his admonition also applies to those who ignore the diversity of homosexuality.

[CHAPTER 4]

Values, Facts, and Disorders

On Homosexuality and Psychiatry

*In the animal kingdom, the rule is, eat or be eaten;
in the human kingdom, define or be defined.*

THOMAS SZASZ

In February 2020, Rabbi Yaakov Ariel received Israel's highest cultural award for his merits as an expert on the Torah. The decision of Israel's minister of education to award the prize to Ariel, however, caused a social media firestorm around the rabbi's earlier remark that homosexuals are "disabled people suffering from a real problem that must be solved with psychological and pharmacological treatments" (quoted in Nachshoni 2016). Despite the backlash, Ariel refused to apologize for pathologizing homosexuality, emphasizing that his attitude toward homosexuality was "compassionate" rather than judgmental because he implied that the condition can be treated and cured.

Unsurprisingly, numerous Israeli LGBTQ organizations urged the minister of education to reconsider his decision, but as yet this has been to no avail. Describing homosexuality as a disability or a disorder (Ariel does not distinguish between them) remains a sign or even a symptom of homonegativity for many, and they may have a point. In many Western societies, homonegative individuals tend to view homosexuality as a disorder. Such views may create more harm for homosexuals, and if we subscribe to the harm principle, then there is a clear moral case to be made against pathologizing homosexuality. In this chapter, we are interested in both the moral and the conceptual case: what, if anything, is morally and theoretically wrong with saying that homosexuality is a disorder?

We begin this chapter by outlining the historical context of the conceptual question, focusing on the psychiatric history of homosexuality in the nineteenth and twentieth centuries. For over 150 years, psychiatrists considered male homosexuality a mental disorder that was part of a series of sexual perversions or deviations such as bestiality, pedophilia, fetishism, and sexual sadism. It was only in the 1970s that homosexuality disappeared from this list. How did psychiatrists come to think that homosexuality was a disorder, and why did they change their minds? We then address the philosophical assumptions and implications of this issue. What do we actually mean when we consider a condition a disorder? What are the criteria that behaviors and mental states should meet in order to be described as symptoms of a disorder? And what do these criteria tell us about homosexuality?

PATHOLOGIZING SEXUAL DEVIANCE

The Western history of human male homosexuality can be described as a history of homonegativity (Sedgwick 2008). The same is true of the history of the gay sciences, with many scientists moralizing and pathologizing homosexuality. (We have, however, already discussed some exceptions, like George Evelyn Hutchinson in the previous chapter.) Earlier, we explored how many nineteenth-century zoologists simply assumed animal homosexuality to be a (mental) disorder. The French entomologist Charles Féré, for example, described homosexual behaviors among insects as "morbid manifestations" and "functional anomalies of the sexual instinct" (Féré 1899, 87). Incidentally, Féré's view coaxed a smile from Karl Heinrich Ulrichs, who joked about zoologists moving to the "Department of Mentally Ill Insects" (Ulrichs 1994, 688). Féré also argued that animal behaviors could not be invoked as proof of the normality of human homosexuality. Human homosexuality, he believed, was as morbid and degenerate as animal homosexuality.

Where and when did this pathologizing originate? It would be tempting to think that its sole source is in psychiatry. As usual, however, history is more complicated, with psychiatrists both echoing and elaborating the disorder view and justifying it against time-honored alternatives.

Historians situate the birth of psychiatry as a science somewhere at the end of the eighteenth century or the beginning of the nineteenth century (Shorter 1997). Barely a few decades later, psychiatrists were already involved in studying all kinds of sexual behaviors, including homosexuality. Still, they were not the first to pathologize sexual deviance. Even before the advent of psychiatry, some physicians had already turned their attention to human sexuality and began to consider sexual deviance as both a cause and an effect of other medical ailments. In the eighteenth century, attempts were made to legitimize Old Testament prohibitions against sexual deviance by referring to the medical risks it posed. In 1761, for example, the physician Samuel-Auguste Tissot published *L'onanisme: Dissertation sur les maladies produites par la masturbation* (*Diseases Caused by Masturbation;*

or, Onanism), and many eighteenth-century Enlightenment thinkers, including Paul-Henry Thiry d'Holbach and Denis Diderot, concurred with Tissot that masturbation, and by extension any form of nonreproductive sex, posed serious health hazards to both body and mind (Gilbert and Barkun 1981; Hare 1962). Although these medical and philosophical writings may not have shaped public opinion in the way nineteenth-century psychiatry did, it is clear that they paved the way for the psychiatric approach to sexual deviance (Cryle and Downing 2009).

So why did the first psychiatrists decide to continue pathologizing sexual deviance? Of course, they had many reasons to do so, and we will discuss some of these reasons as they present themselves in this chapter's historical discussion. Two of them in particular are relevant here, as they take the form of conditions of possibility.

One reason for pathologizing sexual deviance was an increasing political concern with the vitality and health of peoples and nations. Early nineteenth-century politicians worried that degeneration, including sexual deviance, along with falling birth rates would ultimately lead to depopulation and a weakening of the state. To avert these dangers, politicians and other policy makers enlisted the help of psychiatrists and other physicians, whom they considered experts in matters of degeneration. Clearly, the clergy had served its turn as authorities in the sexual domain. Thus it was, for example, that nineteenth-century and early twentieth-century French psychiatrists were paid by the government to take care of the supposedly declining mental hygiene of the French population (Oosterhuis 2000).

Another reason to pathologize sexual deviance was related to the theoretical development of psychiatry as a medical discipline. Initially, psychiatrists considered properly functioning cognition as the hallmark of mental health. Mental disorder, then, was primarily seen as dysfunctional thinking, and sexuality was left largely outside the psychiatric domain. From the 1860s onward, however, psychiatry started broadening its rationalistic concept of "madness" or "insanity" (Berrios 1996). Various psychiatrists proposed new definitions of madness that also encompassed disorders of the will, desires, and emotions, which facilitated a psychiatric account of sexual deviance (Shorter 2008).

Negative attitudes toward unusual sexual behaviors and desires, including moralizing and pathologizing as well as criminalizing them, have existed throughout history, and they continue to exist in many parts of the world. Analytically, these attitudes can easily be differentiated, but historically there are interesting interactions and synergies among these views across different times. One such synergy is the use of the theological term "perversion" in the medical literature up until the 1970s. Another is the use of physicians

Figure 4.1: Musician and composer Frantisek Kotzwara and sex worker Susannah Hill as depicted in *Modern Propensities*, a late eighteenth-century pamphlet on the Kotzwara case, which included Hill's memoirs and a summary of her trial (1797). Source: Anonymous (1797), *Modern Propensities; or, An Essay on the Art of Strangling* (London: J. Dawson).

and psychiatrists as forensic experts (Oosterhuis 2000; Beccalossi 2010), which highlights the influence of the judicial system on the disorder view of homosexuality. In France, one of the most authoritative forensic experts was the physician Auguste Ambroise Tardieu. In his 1857 *Étude médico-légale sur les attentats aux mœurs* (Forensic study on moral offenses), Tardieu noted the importance of listing the various internal and external signs of male homosexuality. Doing so would not only help the legislator and the judge but also provide the French nation with an instrument to efficiently control the morality of these "effeminate" men (Peakman 2009). Throughout the nineteenth century, police forces often called on physicians for guidance on how to deal with sex offenders (Hill 2005). While this sometimes resulted in conflicts between law and medicine, lawyers also actively sought to increase the influence of medical psychologists in court in the hopes of securing the insanity defense (Eigen 1995). One of the oldest and juiciest examples of this trend is the lawsuit that followed the death of the Czech composer and violinist Frantisek Kotzwara in 1792. The lawsuit is described in an anonymous 1797 pamphlet, which contains only one illustration (see fig. 4.1). Kotzwara died while having sex with a sex worker named Susannah Hill. He first asked her to cut off his testicles, which she refused. She did agree, however, to have sex with Kotzwara while he strangled himself. Hill was then prosecuted for murder after his death, but she escaped conviction because her lawyer successfully argued that this kind of sexual scenario was clearly the product of disordered rather than criminal minds (Ober 1984).

SEXUAL INSTINCTS AND THEIR VICISSITUDES

Many eighteenth-century physicians who considered sexual deviance a disorder, rather than a crime or a moral lapse, first focused on the sexual organs, which perhaps explains their obsession with masturbation as the cause of various calamities (A. Davidson 1991). This view gradually changed during the nineteenth century, with physicians increasingly adopting a theoretical framework in which instincts played a central role. There was some debate about exactly how many instincts there were, but most agreed that sexuality was one of them. More particularly, the sexual instinct was considered a reproductive instinct or, relatedly, an instinct for the propagation of the species, and sexual perversions were defined as functional disorders of the sexual instinct. Moreover, there was some consensus that the sexual instinct played a very important role in human life and was susceptible to all sorts of disruptive disturbances (Ellenberger 1970; James 1887).

Interestingly, instinctual theories of sexual deviance were heralded by a watershed event in late eighteenth-century animal science, embodied in the work of the naturalist Georges-Louis Leclerc de Buffon. Throughout the many volumes of his magnum opus *Histoire naturelle*, Buffon managed to convert the age-old Alexandrian rule that sex must be directed toward procreation in order to be moral into a biological mechanism. In Buffon's view, the natural or normal function of sex is to propagate the species. Therefore, those engaging in nonreproductive sex, including homosexuals, are not so much immoral but rather suffering from a dysfunction or abnormality (Brooks 2009).

Phrenology was another mainspring of the new instinctual thinking about sexual deviance (Shortland 1987). One of the pioneers of phrenology, the physiologist Franz Joseph Gall claimed that the brain contained several "organs" or modules related to personality traits and skills. He believed that the human brain had no less than twenty-seven different modules, each responsible for a cluster of functionally coherent behaviors, desires, thoughts, and attitudes. Craniometry, or the measuring of the skull, could help determine the size of these organs in individuals, and their size was thought to be indicative of their talents and weaknesses. One of these measurable organs, Gall claimed, was the instinct for reproduction, which was found in the cerebellum. Gall's theory, then, transformed sexuality into a measurable instinct.

Many psychiatrists quickly adopted some version of the instinct theory of sexual deviance, especially in France under the influence of the famous psychiatrist Jean-Étienne Esquirol. In 1847, Louis Lunier, a student of Esquirol, claimed that the sexual behavior of one of his patients, the necrophiliac sergeant François Bertrand, was due to a pathological failure of the sexual instinct, possibly caused by excessive masturbation (Kamieniak 2003). Some physicians constructed complex classifications of the sexual instinct. Paul Moreau de Tours, for example, argued that the sexual instinct was a sixth sense that could be disturbed in four different ways: it could be too strong, too weak, totally absent, or misdirected. The latter category comprised the sexual perversions, a class of pathologies typified by a deviation of the sexual instinct from its natural aim. This class included homosexuality (A. Davidson 2001).

The instinctual view of the perversions coincided with a change in the relations between the legal and the medical professions with regard to sexual deviance. During the seventeenth and eighteenth centuries, acts against nature, such as homosexuality, were severely punished. In many European countries, men who were caught having anal sex with other men were sentenced to death, provided there were enough witnesses and the act was fully completed ("res in re et effusio seminis"; Gerard and Hekma 1989). Usually, only physical acts were considered illegal, and underlying desires or fantasies were thought to be irrelevant. Still, there are cases on record where both the actions and the associated mental states were condemned. In such cases, it was judged that the sexual criminal was also responsible for his desires. The first psychiatric handbooks subscribed to that view. Until 1860, psychiatrists often treated sexual deviance as the result of a voluntary abandonment of freedom, and therefore individuals suffering from sexual perversions could still be seen as criminals: their actions were the result of a free decision to develop and cultivate disordered desires (Gutmann 2006).

However, this view began to wane after 1860. Increasingly, psychiatrists argued that perverse individuals should be treated and healed rather than punished, in large part because they considered it too simplistic to think that people can actively choose their sexual desires (Waters 2006).

Eager to enhance their professional standing, nineteenth-century psychiatrists believed that they could treat sexual perversions, both medically and psychologically. Their therapies varied according to their etiological hypotheses. Psychiatrists Jean-Martin Charcot and Valentin Magnan, for example, were of the opinion that the psychological origin of sexual abnormalities implied that they were best treated through psychotherapy (Ellenberger 1970). Others recommended hydrotherapy, electrotherapy, and even bloodletting by means of leeches on the penis or the uterus (Oosterhuis 2000).

Despite their differing views on etiology, most psychiatrists subscribed to some version of the degeneration theory, which we discussed in chapter 2. One essential element in most degeneration theories was the conviction that mental disturbances were hereditary disorders and that heredity was destiny. Most degeneration theories also held that aberrant sexual behavior, and particularly masturbation, could trigger or worsen hereditary vulnerability and induce or accelerate a downhill process in which disturbances would progressively and inevitably worsen throughout consecutive generations. Sexual deviance was not just considered to be the *result* of degeneration but also seen as the *cause* of many other "degenerative" or "regressive" illnesses, including alcoholism, pauperism, and moral insanity (Rimke and Hunt 2002).

Nineteenth-century psychiatrists also linked sexual deviations to gender issues. For example, most sexual abnormalities in men were seen as a sign of feminization (Cryle and Downing 2009). Conversely, the criminologist Cesare Lombroso claimed that sexual abnormalities in women were always associated with what he described as "traces of masculinity," such as a low jawline, hard facial features, and a low voice (Lombroso and Ferrero 1999; Seitler 2004). The relationship among gender, degeneration, and deviant sexuality is nowhere more prominent than in psychiatric theories about the nature and origin of homosexuality (Drescher 2010; Groneman 1994). Again, theoreticians considered masturbation to be responsible for the link between feminization and degeneration. After all, the loss of semen was thought to have a detrimental effect on masculinity (Stephens 2008)—a view already defended by Galen in ancient times.

During the 1880s and 1890s, psychiatry's interest in the description, categorization, and etiology of sexual perversions grew rapidly, resulting in many papers and books on the topic. However, historians generally agree that Richard von Krafft-Ebing's taxonomy and discussion of the perversions was by far the most influential of all nineteenth-century contributions to this nascent field.

PSYCHOPATHIA HOMOSEXUALIS

Richard von Krafft-Ebing was a nineteenth-century Austrian forensic psychiatrist. Although he published on many different topics, he is almost solely remembered for his work on sexual deviance. Many consider his *Psychopathia sexualis* as the bible of nineteenth-century sexology (Downing 2010).

Krafft-Ebing was not a radical innovator. Like many of his predecessors, he considered perversions to be functional deviations of the sexual instinct—the *Geslechtstrieb*. He also believed that they were caused by degeneration and that masturbation played an important part in explaining degeneration (Rimke and Hunt 2002). In particular, Krafft-Ebing favored the view that degeneration starts with a hereditary "taintedness" in the family pedigree—a taintedness that leads to an imbalance between the individual's self-control and sexual instincts, which can be aggravated by excessive masturbation (Money 2003; Oosterhuis 2000).

Of course, *Psychopathia sexualis* would never have become one of the most influential books on human sexuality if Krafft-Ebing had limited himself to making rather cautious and unoriginal speculations about the origin of sexual deviance. Then why is the book still seen as a milestone in both psychiatry and sexology?

To start with, the book was the first synthesis of the medical-psychiatric literature on sexual deviance—a synthesis that Krafft-Ebing also illustrated with a whole series of biographical case studies (Oosterhuis 2000). Because of its many case studies and because of Krafft-Ebing's habit of quoting patients extensively, *Psychopathia sexualis* was a medical treatise that explored the boundaries among the medical, the psychological, and—at least in the eyes of many of his contemporaries—the pornographic. The most graphic parts, however, were written in Latin in order to discourage laypeople who were primarily interested in the "pornographic" content (Orrells 2015). Clearly, the book stirred the imagination, and for many it acted as a sort of self-help book avant la lettre. In any case, Krafft-Ebing received many letters from homosexual intellectuals. In these letters, they expressed their gratitude for his work, which, they claimed, offered them genuine insight into their own nature (see fig. 4.2). We will discuss this gratitude in more depth later in this chapter.

On a more theoretical level, Krafft-Ebing's most striking innovation was the idea that sexual abnormalities should be understood not as mental disorders caused by mental states, as Charcot and Magnan believed, but as personality disorders because the perversions constituted an essential part

Figure 4.2: A postcard from Krafft-Ebing's own collection sent to him by an anonymous fellow psychiatrist. It shows a man being "domesticated" by a burlesque woman who is holding reins in her left hand and a whip in her right (1896). © Wellcome Collection, London, L0028609. Creative Commons Public Domain Mark.

of the individual's personality. In other words, Krafft-Ebing saw different perversions as different ways of being a person. The philosopher Arnold Davidson (1991, 314) astutely summarizes this view: "To know a person's sexuality, is to know that person." In the previous chapter, we noted that one of the peculiarities of modern homosexuality is that, for modern homosexuals, sexuality becomes an issue connected to identity. Krafft-Ebing captured (and created)[1] this identity-based feature in his account of male homosexuality in *Psychopathia sexualis*. Later, Michel Foucault (1978, 43) described Krafft-Ebing's revolution, observing that "homosexuality appeared as one of the forms of sexuality when it was transposed from the practice of sodomy into a kind of . . . hermaphroditism of the soul. The sodomite had been a temporary aberration; the homosexual was now a species." If psychiatrists specialized in dealing with the troubled mind, and if homosexuality was a "hermaphroditism of the soul," then psychiatrists were indeed entitled to deal with homosexuality and, by extension, with all kinds of sexual deviance.

The emphasis on a functional and psychological understanding of sexual deviance led Krafft-Ebing to distinguish perversions from perverse acts—acts he called "perversities." Homosexual behavior, for example, is a perversity that differs as such from homosexuality, which is a perversion. The difference between both has to do with their respective locations. Since perversions are functional disorders, their causes are not found in the brain or in the genitals, and they cannot be diagnosed based on behaviors alone. Rather, they are to be found in the person as a whole (A. Davidson 1991; Savoia 2010). According to Krafft-Ebing (1886, 54), "In order to differentiate between disorder (perversion) and vice (perversity) one must investigate the whole personality of the individual and the original motive leading to the perverse act. Therein will be found the key to the diagnosis." Both diagnostically and clinically, the individual's emotions and thoughts were therefore much more important than their sexual behavior. For Krafft-Ebing, someone can be a homosexual without ever putting his fantasies into practice. This "psychologization" further distanced the psychiatric approach to sexuality from the moral and legal approaches. While the latter focused mainly on acts, psychiatry focused on homosexual desires, fantasies, and thoughts (Oosterhuis 2000).

Echoing Moreau de Tours, Krafft-Ebing distinguished among four classes of sexual disturbances, which he jointly called "sexual neuroses": *anesthesia*, or the lessening of the sexual appetite; *hyperesthesia*, or its abnormal increase; *paradoxia*, or its manifestation outside the biologically normal period; and *paraesthesia*, which psychiatry would increasingly focus on and later describe as the sexual perversions or deviations. Individuals in this latter class avoid the procreative act of copulation because they are aroused by inappropriate or unsuitable stimuli. In *Psychopathia sexualis*, Krafft-Ebing subdivided the *paraesthesias* or proper perversions into four large groups: sadism, masochism, fetishism, and contrary sexual instinct (*die konträre Sexualempfindung*)—a disorder that roughly encompassed homosexuality, transvestitism, and transsexuality. Krafft-Ebing added more perversions in later editions of his book, including pedophilia, urolagnia, coprophilia, necrophilia, and various types of bestiality. In the final edition of *Psychopathia sexualis*, the list contained most of the perversions found in twentieth-century and twenty-first-century psychiatric classifications.

An interesting and important consequence of this psychological and functional view of the perversions was that, in principle, homosexual acts and even homosexual desires were not perversions in contexts where no women were available, such as prisons and all-male schools. (In the general introduction, we described this form of homosexuality as situational homosexuality.) In Krafft-Ebing's view, being "gay for the stay" is a perversity, not a perversion, as is heterosexual sadistic bondage, provided it is no obstacle to coitus. Finally, the emphasis on a reproductive criterion helped Krafft-Ebing argue that there was indeed a distinction between the moral and the psychiatric. For example, committing rape is morally reprehensible, but it should not always be seen as a symptom of a disorder, since rape can still serve the biological purpose of reproduction (Hauser 1992).[2] As we will see, similar references to reproduction have remained important in the twentieth century to justify the medicalization of homosexuality.

WAS OEDIPUS GAY?

One cannot overestimate Krafft-Ebing's impact on twentieth-century psychiatry, which includes the revolutionary work of his townsman Sigmund Freud (Hauser 1992). Both men agreed, for example, on the vagueness of the boundaries between sexual perversion and sexual normality, as well as on the importance of sexuality in understanding human personality. Unlike Krafft-Ebing, however, Freud took this last element as the starting point for a comprehensive theory of human nature. He developed a keen interest in the sexual meaning of all sorts of normal and abnormal behaviors and mental states. Ultimately, for Freud, almost all psychopathology was sexual psychopathology (Fedoroff 2009).

Krafft-Ebing's influence can be seen in one of Freud's early theoretical works, *Drei Abhandlungen zur Sexualtheorie* (*Three Essays on the Theory of Sexuality*; Freud 1960b). In the first of these essays, Freud explored the "sexual aberrations" at length, borrowing Krafft-Ebing's distinction between "aberrations according to the sexual aim" and "aberrations according to the sexual object." The latter category included pedophilia, bestiality, and "sexual inversion"—Freud's term for homosexuality. The first category contained oral and anal sex and conditions like fetishism, sadism, masochism, exhibitionism, and voyeurism. This classification seems to suggest that Freud agreed with Krafft-Ebing and other contemporaries in believing that sexual perversions are disorders because they are at variance with the biologically normal object and aim of human sexuality, which is to have reproductive heterosexual intercourse.

On closer inspection, however, Freud's approach was quite different from the psychiatric consensus of the time. In *Three Essays*, Freud was not interested in the perverse disposition that kept individuals from desiring normal sex. His project was rather to give an account of human sexuality as such—a sexuality he considered intrinsically perverse: "The disposition to perversions is itself of no great rarity but must form a part of what passes as the normal constitution" (Freud 1960b, 171). Lacking a natural or innate link between sexuality and its objects and aims, all humans grow up as "polymorphously perverse" beings. In Freud's view, this perversity reveals itself most clearly in childhood, when children play all sorts of sexual games, urged on by partial sexual instincts, including homosexuality. In many people, however, these sexual instincts continue throughout their adulthood.

Like Krafft-Ebing and many other contemporaries, Freud used instinct theory to make sense of the perversions. In Freud's hands, however, the perversions transformed from deviations of a single, normal sexual instinct to instincts or urges themselves (A. Davidson 1987). While this change of perspective further blurred the line between normal and abnormal sexuality, Freud never erased that line altogether. In fact, he strongly believed that the child's polymorphously perverse sexuality could give rise to both healthy and pathological adult sexual fantasies and behaviors. In his early works, including *Three Essays*, he focused on exclusivity as the hallmark of deviance (Savoia 2010). Masochism is a disorder not because one enjoys being humiliated or punished but rather because one *only* enjoys humiliation and punishment, without any interest in coitus or other sexual activities. (Interestingly, Freud's emphasis on the exclusivity criterion, together with his claim that humans are not biologically programmed for heterosexual intercourse, seems to imply that both exclusive homosexuality and heterosexuality are perversions. Freud himself never drew that conclusion.)[3]

In *Three Essays*, Freud considered fixation to be the most important pathogenic mechanism. Fixation came about after one found too much pleasure in a particular perverse childhood activity, so that when puberty arrived, along with its hormonal surge, one sexually "regressed" to that perverse source of infantile sexual pleasure. This general scheme was supplemented with many specific details for each perversion. For example, Freud explained Leonardo da Vinci's homosexuality by claiming that he longed to return to the period in his childhood in which he experienced (too) much loving care from his mother. According to Freud, da Vinci's homosexual desires simply repeated his mother's love for him. Da Vinci identified with his mother to regain this lost pleasure: he preferred boys, much like (and because) his mother had preferred him in his early childhood.

From 1915 onward, Freud began to adjust his views on the etiology of the perversions to accommodate the many new concepts he had added to his theory since the first publication of *Three Essays*—concepts such as penis envy, castration anxiety, and narcissism. Initially, he only made minor modifications, but in 1927 he introduced a dramatic revision of his theory on sexual perversions. While still reserving an important role for fixation, which he originally related to physiological causes, he now added another cause: the denial of a traumatizing sexual experience, particularly the castration anxiety that accompanied a child's inevitable oedipal desire (Freud 1964b). To be able to experience sexual pleasure, masochists sexualize castration, while fetishists sexualize those objects that help them keep the fear of castration at bay. Freud's belief that all male perversions could be understood as a defense against castration anxiety had a huge impact on later psychoanalytic theories about sexual perversions (Metzl 2004).

One of the surprising consequences of this theoretical turn is that Freud no longer considered homosexuality and some other sexual perversions as disorders. In a now famous 1935 letter to the mother of a homosexual young man, he wrote: "Homosexuality is assuredly no advantage, but it is nothing to be ashamed of, no vice, no degradation; it cannot be classified as an illness; we consider it to be a variation of the sexual function, produced by a certain arrest of sexual development" (Freud 1960a, 423). Freud still considered homosexuality to be the result of fixation ("a certain arrest of sexual development"), but he no longer saw it is as a disorder or a perversion since homosexuality was not associated with oedipal conflicts or castration anxiety. Unlike real perversions, homosexuality should be understood not as an adaptation to or a defense against something unpleasant (fear, pain, anger) but as the result of too much pleasure during the homosexual (or "narcissistic") phase of individual development.

HOMOSEXUALITY IN EARLY SEXOLOGY

The birth of psychoanalysis more or less coincided with the birth of sexology as a scientific discipline. Albert Moll's *Untersuchungen über die Libido sexualis* (*Libido sexualis: Studies in the Psychosexual Laws of Love*), mentioned in chapter 2, was published in 1897, as was Henry Havelock Ellis's first volume of the *Studies in the Psychology of Sex*. Less than two years later, Magnus Hirschfeld started the first specialized yearbook *Jahrbuch für sexuelle Zwischenstufen* (Yearbook for sexual intermediaries), in which he tried to cover the steady growth of the scientific literature on sexuality.

Freud's work was in constant dialogue and debate with these early sexologists. The central message of Ellis's work, for example, was that sexual interests, like many other biological properties, are normally distributed across a bell curve, with most of the population located near the mean—sexual normality—and only a handful of individuals located at either extreme. This message fit well with Freud's dimensional view of human sexuality (Fedoroff 2009). However, Freud and the early sexologists also had their disagreements, one of which was about the etiology of unusual sexual interests. In contradiction to Freud, Ellis and Hirschfeld claimed that one's personal history or development did not affect one's sexual orientation. They instead considered such orientation to be innate (Crozier 2008). Another disagreement related to their objectives. Even though Freud himself remained skeptical about the possibility of curing the sexually perverted (Freud 1964a), psychoanalysis presented itself mainly as a method of medical treatment, and later on some psychoanalysts certainly capitalized on such treatments, including conversion therapy, for homosexuals. The early sexologists, by contrast, were not interested in treating sexual disorders. Their main objective was to describe and classify the many variants of human sexuality. If anything, many of these sexologists wanted to change society rather than cure the individual, and their objective was thus more emancipatory than therapeutic. Hirschfeld even defined sexology as a progressive science, since it revealed that the so-called deviations from the sexual norm were neither pathological nor dangerous to society (Meyenburg and Sigusch 1977). As we noted in chapter 1, his motto was *Per scientiam ad justitiam*—From science to justice (see fig. 1.1).[4]

Central to the sexological project was the view that perversions were not as problematic as many psychiatrists and educators had claimed. The psychiatrist (and dermatologist) Iwan Bloch, for example, argued that it was important to approach sexuality not just from a medical perspective. Anthropological and historical studies were also needed to get a complete and nuanced picture of human sexuality (Matte 2005). Such studies showed that many of the so-called sexual perversions occurred in all cultures and times; what really needed to be explained, according to Bloch, was not why they existed but why people thought they should be eradicated.

Although Bloch, Hirschfeld, and Ellis still held on to the view that the sexual instinct was primarily a procreative instinct, they also emphasized that many people enjoy nonprocreative sex as an important and quite harmless source of pleasure. Most of these early sexologists spent their intellectual efforts studying the causes and the natural history of homosexuality, as the most visible and prevalent of all perversions, and their activism centered on reforming the laws concerning homosexuality (Crozier 2008; Hirschfeld 1952). In fact, they shared this objective with many psychiatrists of their time. Even psychiatrists who considered homosexuality a disorder, like Emil Kraepelin and Ernst Rüdin (Franz Kallmann's mentor), demanded the abolition of paragraph 175, a provision of the German criminal code that criminalized homosexual acts between men (Mildenberger 2007).

Ellis's books and ideas found a generally receptive audience in the United States, perhaps because his work was less overtly theoretical than the German and French approaches to sexuality and sexual deviance. Even before World War I, American sexology mainly focused on empirical and statistical studies, often based on extensive questionnaires. These studies of the sex lives of ordinary people were intended as support for what was seen as social amelioration, even though as a whole, American sexology (initially) did not have the same emancipatory aims as its German and British counterparts (Waters 2006).

To some extent, Alfred Kinsey's work belongs to this American tradition of statistical research on sexual behavior.[5] Kinsey, a biologist, won fame in postwar America with his *Sexual Behavior in the Human Male* (Kinsey, Martin, and Pomeroy 1948) and *Sexual Behavior in the Human Female* (Kinsey et al. 1953). In both works, he combined a statistical approach of case histories with a biological view of human sexual practices, and he used this biostatistical framework to argue, much like some of his European colleagues, that most so-called sexual perversions were not disorders at all (Meyerowitz 2001). First, he believed that his reports indicated that many supposedly deviant sexual practices were actually quite common in the American population. Half the men in his data admitted to being sexually aroused by another man at least once in their life. Second, it made no sense to him to assume that these conditions violate some natural norm, as most of the perversions, including homosexuality, can also be found in nonhuman animals.

Overall, and rather unfortunately, both European and American early sexologists had only a limited influence on twentieth-century psychiatric theorizing about human sexuality (Bullough 1998). This may be due to their lack of interest in treating or "curing" individual sexual desires and behaviors, which clearly distanced them from Freud's followers who, as we will see shortly, were only too happy to extend their therapeutic program to sexual deviance. In doing so, psychoanalysis had an enormous impact on twentieth-century conceptualizations of sexual deviance (Shorter 1997). That said, the evidence and the arguments of the early sexologists did play an important role in the debate about how to distinguish between normal and abnormal sexuality—an issue that haunted psychiatry throughout the twentieth century, reaching a climax in the 1970s.

In the next sections, we discuss psychiatry's dealings with sexual deviance in the second half of the twentieth century, focusing on the place and role of homosexuality in the consecutive editions of the American Psychiatric Association's *Diagnostic and Statistical Manual of Mental Disorders*. We will argue that the debate about homosexuality has significantly influenced the development of this mental health manual, as it was the driving force behind the formulation of a general definition of the concept of mental disorder. It will also become clear, however, that later editions of the manual had difficulty complying with that definition, thus creating confusion about the place of sexual perversions in psychiatric classification.

THE EARLY *DSM* AND THE HOMOSEXUALITY CONTROVERSY

The American Psychiatric Association (APA) is probably the world's most powerful professional organization of psychiatrists. It is involved in the publication of books and journals and the organization of various health campaigns and conferences. Its most visible work, however, is the preparation and publication of the *Diagnostic and Statistical Manual of Mental Disorders* (*DSM*). Despite persistent and sharp criticism from many different angles, the *DSM* is still the leading clinical diagnostic manual today, and it is also widely used for administrative and research purposes.

The *DSM* originated from the need for a uniform method of reporting statistics from the many mental hospitals in early twentieth-century America (Grob 1991). Its predecessor, the *Statistical Manual for the Use of Hospitals for Mental Diseases*, first published in 1918, reflected the population of these hospitals and concentrated mostly on severe brain disorders, often with an organic etiology. One of the manual's clinical groups was given the enigmatic name "Not Insane," and it included a category called "Constitutional Psychopathic Personality (without Psychosis)," which referred to "criminal traits, moral deficiency, tramp life, sexual perversions and various temperamental peculiarities" (National Committee for Mental Hygiene 1918, 27). In a way, the *DSM*'s predecessor did not consider sexual perversions as genuine mental disorders. Its message was more ambiguous, however, since the manual also referred to "perverts," tramps, and criminals as "pathological" and even "abnormal personalities."

The *Statistical Manual*'s original mission to collect mental hospital data soon encountered unanticipated difficulties. American soldiers returned from the strains and rigors at the fronts of World War II with illnesses that were nowhere to be found in the manual. Combat fatigue and shell shock produced relatively mild mental disorders, at least when compared to the gravely afflicted patients kept in mental hospitals. Faced with an enormous new patient population, the APA quickly understood the need to expand its stock of disorder categories. In 1952, it published the *Diagnostic and Statistical Manual: Mental Disorders* (*DSM-I*; APA 1952).

DSM-I had very little to say about sexual deviations. It catalogued them as one of the "Sociopathic Personality Disturbances" that were part of the general category of personality disorders. Interestingly, in the description of sociopathic personality disturbances, it notes that "individuals to be placed in this category are ill primarily in terms of society and of conformity with the prevailing cultural milieu, and not only in terms of personal discomfort and relations with other individuals" (APA 1952, 38). It is perhaps no coincidence that this analysis was published around the same time as Kinsey's *Sexual Behavior in the Human Male*, where he claims that "the problem of the so-called sexual perversions is not so much one of psychopathology as it is a matter of adjustment between an individual and the society in which he lives" (Kinsey, Martin, and Pomeroy 1948, 32). *DSM-I*'s description also resonates with the work of the early European sexologists, and it is one of the rare occasions where the editors of *DSM-I* hint at a definition of mental disorder. The first *DSM* did not provide such a definition, at least not explicitly, and neither did *DSM-II* (Cooper 2005). However, their general outlook suggested that mental illness should be understood either in terms of some organic defect, as in the case of the many brain disorders listed, and/or in terms of personal distress, as in the case of the neuroses. Perversions fell outside this implicit definition of mental illness, as they were seen primarily as instances of social deviance rather than mental illness.

Like any psychiatric classification, *DSM-I* was a product of its time. Its descriptions of disorder categories were riddled with psychoanalytic terms and concepts, such as "unconscious internal conflict" (APA 1952, 47), "projection mechanism," and "regressive reaction" (36). Contrary to what some historians of psychiatry (e.g., Shorter 1997) and biological psychiatrists (e.g., Maxmen 1985) have claimed, the second edition of the *DSM*, first published in 1968, did not really continue this tradition. Its descriptions were shorter, and speculations as to the causes and mechanisms of disorders were kept to a minimum. The pursuit of a theory-neutral or atheoretical nomenclature would become increasingly important in later editions of the *DSM*.

As to the perversions, one of the novelties of *DSM-II* was the introduction of an extensive list of eight sexual deviations, all of which were listed in the final edition of Krafft-Ebing's *Psychopathia sexualis*: homosexuality, fetishism, pedophilia, transvestitism, exhibitionism, voyeurism, sadism, and masochism.

In addition, while *DSM-I* and its precursor considered the perversions as a kind of personality disturbance, *DSM-II* listed them under the rather vague heading of "Certain Non-Psychotic Mental Disorders." More importantly, *DSM-II* omitted all references to the pathogenic power of social norms in the general description of the sexual deviations. This edition resolutely focused on the personal distress accompanying these deviations: "Even though many find their practices distasteful, they remain unable to substitute normal sexual behavior for them" (APA 1968, 44). Much like the concern for theory neutrality, the increasing emphasis on the criterion of personal distress was an early pronouncement of the looming landslide that was to follow the publication of *DSM-III*.

The 1970s were turbulent times for the APA. Since World War II, the majority of its members had been practicing psychoanalysts, but now the powers of psychoanalysis were waning (Decker 2007). The decline of psychoanalysis set the stage for a new wave of biomedical, research-oriented psychiatrists, which created a power struggle within the APA—a struggle that culminated in one of the most pressing, and perhaps most embarrassing, issues in the build-up to the making of *DSM-III*: the problem of homosexuality.

Unlike *DSM-I*, *DSM-II* unambiguously qualified homosexuality as a mental disorder. Many commentators have associated this view with the predominance of psychoanalysis in the early postwar intellectual climate (e.g., Friedman and Downey 1998). A large number of psychoanalysts disagreed with Freud on this topic, who, as we noted earlier, did not consider all homosexuals mentally ill. Another important difference between Freud and mid-twentieth-century American psychoanalysts was their views on the need for, and the prospects of, therapeutic intervention. Freud was remarkably clear on this topic: "In general to undertake to convert a fully developed homosexual into a heterosexual is not much more promising than to do the reverse, only that for good practical reasons the latter is never attempted" (Freud 1955, 32). The therapeutic optimism of postwar psychoanalytic psychiatrists, however, was markedly greater than Freud's, and many of them were actively engaged in so-called conversion therapy when the controversy over homosexuality erupted in the early 1970s (see, e.g., Bieber et al. 1962; and see chapter 1).

Postwar psychoanalysts firmly adhered to the disorder view of homosexuality because they considered heterosexuality to be a natural norm. In the words of one psychoanalyst: "Humans are biologically programmed for heterosexuality" (Bieber 1987, 425). Traumatizing experiences and disturbed parent-child or peer relationships were thought to dislocate this "natural urge" and cause abnormal sexual behavior. Later in this chapter, we will see that psychoanalysts were not the only ones to defend the disorder view in the 1960s and 1970s. The American philosopher Christopher Boorse also conceptualized homosexuality as a disorder because it conflicts with "one normal function of sexual desire," which is "to promote reproduction" (Boorse 1975, 63).

Throughout the 1960s, however, the disorder view came under increasing attack from a variety of actors, including gay activists and public intellectuals. According to Judd Marmor, an outspoken opponent of the psychoanalytic view, "It is our task as psychiatrists to be healers of the distressed, not watchdogs of our social mores" (Marmor 1973, 1209). Critics of the disorder view put forward a number of arguments, some of which were reminiscent of the work of early European and American sexologists, including Hirschfeld and Ellis. First, they claimed that homosexuality was biologically natural. Marmor, for example, paraphrased "an eminent biologist" and said that "human homosexuality reflects the essential bisexual character of our mammalian inheritance" (1209).[6] Second, they argued that even if heterosexuality was a natural norm, it did not necessarily follow that homosexuality was a disorder. Celibacy and vegetarianism can also be considered as "violations" of a natural norm, Marmor argued, yet these are not generally considered disorders. Third, they provided some evidence that not all homosexuals are, or were, ill. Most of the evidence brought forward by psychoanalysts came from clinical practice, and critics argued that it was obvious that such evidence could not represent the homosexual population as a whole (Fuller Torrey 1974). Finally, even if the overwhelming majority of contemporary homosexuals turned out to have psychological problems, the question would be whether they do so because of some inherent pathology, as psychoanalysts maintained, or because of the oppressive power of a still vehemently homonegative society (Gold 1973).

By devising arguments that showed that homosexuality was neither abnormal nor a disorder, critics of the disorder view fueled the work of gay activist groups. From 1970 onward, some of these groups started protesting the annual meetings of the APA, where leading psychoanalysts presented their evidence for the disorder view (Bayer 1981). In the midst of this dispute between activists and psychoanalysts, the young psychiatrist Robert Spitzer stepped up as a go-between. Spitzer was originally convinced that homosexuality did belong in the *DSM*. Various events, however, including his attendance at an informal meeting of the "Gay-PA"—a secret group of homosexual APA members later known as the Association of Gay and Lesbian Psychiatrists—made him realize that many homosexuals were actually healthy and high-functioning individuals who were often satisfied with their sexuality (Bayer 1987). Soon afterward, he drafted a compromise: homosexuality as such was to be removed from the *DSM* and replaced by "Sexual Orientation Disturbance," which referred only to individuals who were troubled by their own sexual orientation.

One of the important motivations for this proposal was an attempt to define the concept of mental disorder. In Spitzer's view, such a definition should entail two elements: "It must either regularly cause subjective distress, or regularly be associated with some generalized impairment in social effectiveness or functioning" (Spitzer 1973, 1215). Many homosexuals did not fulfill either of these criteria, and therefore they should not be considered mentally ill. Importantly, Spitzer did not see homosexuality as normal either: "No doubt, homosexual activist groups will claim that psychiatry has at last recognized that homosexuality is as 'normal' as heterosexuality. They will be wrong" (1216). To meet the expected objections of the psychoanalytic community, he suggested describing homosexuality as an "irregular form of sexual development" that is "suboptimal" when compared to heterosexuality (1215). Yet suboptimal behavior, he argued, need not necessarily constitute disorder, such as in the examples of celibacy, racism, religious fanaticism, or vegetarianism, which he jokingly described as an "unnatural avoidance of carnivorous behavior" (1215; see also Spitzer 1981).

Despite its obvious diplomatic qualities, Spitzer's proposal was met with fierce protest, and for different reasons. Activists were angry that homosexuality was not considered as valuable as heterosexuality, while psychoanalysts repeatedly called on APA officials not to capitulate to political pressure. Nevertheless, in December 1973 the APA's board of trustees unanimously accepted the proposal to eliminate homosexuality from the *DSM* and to include the new category of "Sexual Orientation Disturbance." Following further protest from a number of leading psychoanalysts, the APA organized a referendum: Should homosexuality be in the APA nomenclature or not? The APA membership accepted Spitzer's proposal by 58 percent, and consequently homosexuality as such was removed from the seventh printing of *DSM-II* in 1974.

According to some commentators, the referendum was a public relations disaster for the APA, and devising a psychiatric nomenclature turned into a matter of politics rather than science (Shorter 1997; Kirk and Kutchins 1992). In our view, this judgment is too harsh. After all, the vote arose from a series of debates within various scientific bodies, all of which interpreted and weighed scientific evidence. The referendum only came about because all parties involved were convinced that science was on their side (Stoller et al. 1973). Spitzer was actually one of the few who realized that, in this debate, science could not possibly have the last word. Part of the problem, he believed, was that the concept of disorder was not value neutral (Spitzer 1981).

Despite this sobering history, many of the architects of the *DSM* continued (and continue) to claim that the manual, and especially its third edition, was the first real evidence-based and scientifically sound psychiatric classification. J. Maxmen, one of the proponents of the new approach, noted, "The old psychiatry derives from theory, the new psychiatry from fact" (Maxmen 1985, 31). For a long time, Spitzer also stood by such views (see, e.g., Spitzer 2001). For some reason, however, he changed his mind a couple of years before his death in 2015. In an interview from early 2007, he conceded that the *DSM-III* task force did not always rely on research evidence. The following is from a conversation between Spitzer and his colleague Max Fink about how new disorder categories were included in the nomenclature:

> SPITZER: You have to have a lobby, that's how. You have to have troops.
>
> FINK: So it's not a matter of...
>
> SPITZER: Having the data? No.
>
> FINK: It's nothing to do with science then, and nothing to do with evidence?
>
> Spitzer nodded. (Shorter 2008, 168)

The interviewers seem to have been shocked at this "confession," but Spitzer's honesty should not be too surprising. Immediately after the APA board's decision to remove homosexuality from their manual, the psychoanalyst Irving Bieber publicly asked Spitzer whether he would consider deleting other sexual deviations from the *DSM* too. Spitzer answered: "I haven't given much thought to [these problems] and perhaps that is because the voyeurs and the fetishists [unlike the homosexuals] have not yet organized themselves and forced us to do that" (quoted in Bayer 1987, 397; see also Bieber 1987, 433).

THE UNHAPPY HOMOSEXUAL AND THE HAPPY FETISHIST

In May 1974, immediately after the referendum, the APA appointed Spitzer as the new chair of the Task Force on Nomenclature and Statistics, a committee within the APA that made the final decisions on psychiatric nomenclature. One of Spitzer's first decisions was to remove all psychoanalysts from his task force. Consequently, *DSM-III*, first published in 1980, differed dramatically from *DSM-II*. One novelty was the introduction of diagnostic criteria—in order to be eligible for a particular diagnosis, the patient had to fulfill a specific number of such criteria. Together with a significant increase in the number of disorder categories, the inclusion of these criteria more than doubled the size of the manual's previous edition. Another interesting change was an attempt, on the very first pages of the manual, to define the concept of mental disorder: "In DSM-III each of the mental disorders is conceptualized as a clinically significant behavioral or psychological syndrome or pattern that occurs in an individual and that is typically associated with either a painful symptom (distress) or impairment in one or more important areas of functioning (disability). In addition, there is an inference that there is a behavioral, psychological, or biological dysfunction, and that the disturbance is not only in the relationship between the individual and society" (APA 1980, 6).

As we explained earlier, the APA removed homosexuality from *DSM-II* mainly because it did not fit this definition of mental disorder that, according to Spitzer (1981), was also implicitly employed when constructing the first two editions of the manual. This implicit definition was based on two criteria: distress and disability (or functional impairment). Because many homosexuals were not in any way distressed by their sexual orientation, and since most of them appeared to function very well, both socially and professionally, it was clear that homosexuality should be excluded from the manual. But what about other sexual deviations, such as voyeurism or sexual sadism? What evidence was there to believe that these conditions, unlike homosexuality, did cause significant distress or disability?

Spitzer himself believed that the inclusion of some perversions, particularly voyeurism and fetishism, as mental disorders was "questionable," and he was aware that many similarly expected him, following the APA decision about homosexuality, to remove these conditions from the manual (Spitzer 1981, 406). Still, all of *DSM-II*'s sexual deviations reappeared in *DSM-III*, if only under a different name ("Paraphilias") and (again) in a different diagnostic class ("Psychosexual Disorders"). The term "paraphilias" was preferred to the old "sexual deviations" in emphasizing that "the deviation (para) is in that to which the individual is attracted (philia)" (APA 1980, 267). Not only was the new name more accurate, however; it also had the bonus of sounding more scientific and less moralistic or judgmental (Bullough 2003). The manual listed as paraphilias the usual suspects: fetishism, transvestitism, zoophilia, pedophilia, exhibitionism, voyeurism, sexual masochism, sexual sadism, and some "atypical" paraphilias, such as frotteurism and necrophilia.

So why did *DSM-III* continue to present these conditions as mental disorders? Three reasons stand out. First, paraphilias are disorders because they require "unusual or bizarre imagery or acts . . . [as] necessary for sexual excitement." Second, the objects and situations characteristic for these conditions "are not part of normative arousal-activity patterns." Third, these patterns "in varying degrees may interfere with the capacity for reciprocal sexual activity" (APA 1980, 261). The first two criteria seem to say only that these preferences are medically abnormal because they are statistically abnormal, but from a merely statistical point of view, homosexuality is also a disorder. Generally, statistical abnormality is not a good enough reason to consider something a medical problem. Having green eyes, for example, is statistically abnormal, as it occurs in less than 3 percent of the world's population, but it would be absurd to say that being green-eyed is a disorder. Only the third criterion is a vaguely legitimate reason to exclude homosexuality, even though one may ask whether it really applies to all paraphilias and whether it is a scientific or a moral criterion. In fact, up to the 1970s, many psychoanalytic psychiatrists considered homosexuality as a kind of sexuality that hinders "true interaction" with others, because of its allegedly narcissistic origin (see, e.g., Duyckaerts 1966).

DSM-III explicitly acknowledged that paraphilic fantasies and actions can be part of a healthy sexual relationship. Many men find lingerie exciting, as do fetishists, and elements of sexual sadism and masochism have a place in the bedrooms and kitchens of many healthy individuals (APA 1980). Therefore, *DSM-III* assumed a continuum between normal and disordered sexual desires, fantasies, and actions. Still, in order to be able to draw a line between sexual health and sexual deviance, the manual introduced "Diagnostic Criterion A," which stipulated that unusual sexual fantasies or actions should only be considered as symptoms of a paraphilia when they become "insistently and involuntarily repetitive," "repeatedly preferred or exclusive," or even "necessary" to achieve sexual gratification (APA 1980, 267, 273–74). In a way, this returned—presumably unintentionally—the antipsychoanalytic *DSM-III* to early Freudian thinking about sexual deviance. More importantly, the general definition of mental disorder in *DSM-III* did not mention either exclusivity or repetition, while the manual did use that definition to legitimate the removal of homosexuality. Homosexuality did not disappear from *DSM-II* because homosexual men and women could enjoy heterosexual fantasies and actions from time to time (criterion A) but because many homosexuals were not (necessarily) distressed or impaired by their sexual orientation. Finally, it was absolutely unclear how the exclusivity criterion related to the other criteria of distress and impairment.

As mentioned before, the removal of homosexuality from the *DSM* was a gradual process. Spitzer's 1973 proposal suggested replacing homosexuality (in one of the last prints of *DSM-II*) with a new disorder category, "Sexual Orientation Disturbance," to refer to those individuals who are distressed or impaired by their sexual orientation. In a way, this provided psychoanalysts with new ammunition to ridicule the APA's decisions about homosexuality. In their view, the new category implied that, in principle, heterosexuality could also be considered a disorder, at least if it was experienced as distressing or impairing. To answer this critique, *DSM-III* replaced "Sexual Orientation Disturbance" with "Ego-Dystonic Homosexuality," but the underlying idea remained the same: homosexuality is only a disorder when accompanied by distress or impairment. Therein lay the main difference between homosexuality and the paraphilias: the latter were disorders regardless of the suffering they may cause in the individual.

Homosexuality finally disappeared from the *DSM* altogether when "Ego-Dystonic Homosexuality" was dropped in a revision of *DSM-III* published in 1987 as *DSM-III-R*. Now homosexuality was officially no longer a disorder, even if it came with feelings of distress. Regarding the paraphilias, considerations of exclusivity and repetition were no longer deemed essential. They were replaced by two basic diagnostic criteria that applied to all the listed paraphilias. Criterion A required the presence of "recurrent intense sexual urges and sexually arousing fantasies, over a period of at least six months," while criterion B stipulated that "the person has acted on these urges, *or* is markedly distressed by them" (APA 1987, 282–90; italics ours). The note about distress in criterion B could be seen as an attempt to fit the paraphilias into the manual's general definition of mental disorder. Curiously, however, there was no mention of impairment or disability, while a new criterion was added rather surreptitiously: according to *DSM-III-R*, some urges and fantasies needed only to be *acted on* to indicate disorder, even if they did not cause any distress to the individual. Let us call this new criterion "the action criterion."

In a kind mood, one could argue that the action criterion fits within a broadly interpreted impairment criterion. If someone's sexual actions harm others, there are good reasons to assume that there is something wrong with that individual's social functioning. That said, the action criterion is problematic for at least two reasons. First, it is at odds with an intellectual heritage that dates back to the work of Krafft-Ebing and his distinction between sexual perversities and sexual perversions.[7] Second, and perhaps most important, there is no mention of the action criterion in the *DSM*'s general definition of mental disorder. In less than a decade, the APA managed to come up with two ad hoc disorder criteria to warrant its decision to keep the paraphilias listed in the *DSM*, and in the process the association ignored its own general definition of mental disorder, which was used to remove homosexuality.

In *DSM-IV*, published in 1994, the action criterion was traded in and replaced by the impairment criterion, and diagnostic criterion B was reformulated to require that "the [paraphilic] fantasies, sexual urges, or behaviors cause clinically significant distress or impairment in social, occupational, or other important areas of functioning" (APA 1994, 523). Absent distress or impairment, unusual sexual fantasies, urges, or behaviors were considered nonpathological. One may condemn them as instances of criminality or eccentricity but not as disorders. While this wording was by far the most consistent vis-à-vis the *DSM*'s own definition of disorder, the amendment was short-lived (Moser and Kleinplatz 2005). Pressure from conservative lobbyists and some psychiatrists led the editors of *DSM-IV-TR* to revert to *DSM-III-R*'s criterion B, which required either acting on unusual sexual urges or fantasies *or* experiencing distress about them (APA 2000, 566).

For some years, the APA was praised for its decision to remove homosexuality from the *DSM*. The very same decision, however, brought the APA to bay, as it struggled and ultimately failed to explain why it continued to pathologize other sexual deviations or paraphilias. This failure has primarily been caused by the manual's inconsistent application of its own definition of disorder in dealing with sexual deviance. All too often, it invokes ad hoc criteria, such as the exclusivity criterion and the action criterion, which makes it more difficult to determine whether the *DSM* can help distinguish between normal and abnormal sexuality (or, more generally, between healthy and disordered variation in human thinking and behavior).

In the remaining sections of this chapter, we explore two solutions to this problem. A first solution attempts to refine or elucidate the current *DSM* definition of disorder, or at least apply it more consistently. While this solution may solve some issues, we think it runs into its own problems. After all, the *DSM* definition is not philosophically sophisticated, and attempts to make it more sophisticated risk creating confusion about the disorder status of homosexuality. A second solution urges us to ditch the *DSM* definition of disorder and start looking for a better one. We prefer this solution, and although we do not propose our own definition of mental disorder, we will describe how homosexuality can play a central role in the definitional strategy that should be followed to produce such a definition.

In our general introduction, we mentioned that philosophers can be useful in definitional debates, so we will now gradually change our tack from the history to the philosophy of science. We will discuss how philosophers have tried to define mental disorder, how these philosophical definitions fit within a philosophical approach called conceptual analysis, and what the discussion about the disorder view of homosexuality can tell us about the shortcomings of conceptual analysis.

A VICTORY FOR SCIENCE?

The psychiatrists Michael First and Allen Frances were responsible for editing *DSM-III-R* and introducing the action criterion. Paraphiliacs, they argued, suffer from a disorder as soon as they move away from fantasies to activities with nonconsenting individuals. Such activities suffice to pathologize them. Later, in a 2008 editorial, First and Frances reluctantly admitted that the action criterion blurs the line between disorder and ordinary criminality (First and Frances 2008, 1240; see also Gert and Culver 2009). While they acknowledged the importance of that line, they did not know how to draw it.

A few years later, First renewed the attempt to justify scientifically the distinction between disordered and criminal or morally undesirable activities. To avoid confusion, he argued, it is essential to take into account the nature of the fantasies and urges that precede or accompany the offenses (First 2010, 1240). First's recommendation was to revive a forgotten aspect of the *DSM*'s general definition of mental disorder. The definition specifies that a condition can only qualify as a disorder if it causes distress or impairment and if it is considered "a manifestation of a behavioral, psychological, or biological dysfunction in the individual" (APA 2000, xxiv). He failed to specify, however, how to define and operationalize the concept of "dysfunction." In fact, no edition of the *DSM* provides such a definition, which means they are as useless in defining disorder as Krafft-Ebing's speculations about an underlying "general neuropathic or psychopathic condition" (Krafft-Ebing 1886, 501).

First's appeal to dysfunction is reminiscent of the belief that scientifically documented facts can help us distinguish between mental health and mental disorder. Philosophers consider this belief to be the core of the naturalist position in the debate about the nature of disorder. Naturalists argue that disorders are simply biological dysfunctions. Normativists, by contrast, argue that disorders are disvalued rather than dysfunctional states, in the sense that they are in some way harmful or undesirable. By definition, according to normativists, a disorder is a bad thing to have.

Naturalism played an important role in the controversy over homosexuality in the 1970s. Most of the participants in the debate provided (what they saw as) scientific evidence to prove their point. In an address given at the University of California, Irvine, in 2002, Marmor stated that the decision to remove homosexuality from the *DSM* was determined by science alone: "The fact is that the decision to remove homosexuality from the *DSM* was *not* based on gay political pressure but on scientific correctness, and only after a full year of exploratory hearings and study of the issue by the APA's Committee on Nomenclature" (quoted in Drescher and Merlino 2007, 86; italics in original). Marmor knew that supporters of the disorder view of homosexuality also claimed to be naturalists and that they argued that their judgments were based on facts. According to Marmor, however, they were "normativists in disguise" who considered homosexuality a disorder because they disapproved of homosexuality.

Marmor firmly believed that true naturalism was at odds with the disorder view of homosexuality. However, in the 1970s and 1980s, supporters of the disorder view included not only psychoanalysts like Bieber and Charles Socarides but also a number of naturalist philosophers. In 1984, for example, the controversial[8] philosopher Michael Levin published a paper in *The Monist*, a high-profile philosophical journal, in which he argued that homosexuality is medically abnormal and that its abnormality is a scientific fact. According to Levin, everybody would agree that if someone pulled all their teeth to make a necklace, then the individual is not using the teeth as nature intended. Such an action would rightly be considered abnormal. Similarly, if a homosexual man uses his penis for homosexual acts (Levin focuses on anal sex), then he is not using his penis as intended. Levin further argues that because natural selection tends to link positive feelings to the correct or natural use of body parts, we can expect those who misuse their body parts to be less happy than those who use them correctly. People who use their teeth to chew tend to be happier than those who use their teeth as a necklace. Likewise, those who insert their penises into anuses tend to be less happy than those who insert them into vaginas. Levin (1984, 261) also notes that those "proto-human males who enjoyed inserting their penises into each other's anuses have left no descendants." He acknowledges that some people may still enjoy homosexual activities but emphasizes that recent surveys revealed that homosexuals are on average much less happy than heterosexuals, even in more tolerant societies like Denmark or the Netherlands. Their unhappiness, according to Levin, was due to their "unnatural" use of body parts.

Even though we subscribe to Levin's premise that natural selection tends to associate positive feelings with the evolved use of body parts, there are still numerous examples where "misuse" does not result in less happiness. Our fingers did not evolve to play the piano, but many piano players are happy.

Levin was not the only philosopher to make a naturalist case for the disorder view. Earlier we mentioned Christopher Boorse, who first defended his biostatistical analysis in the mid-1970s. To this day, Boorse's account is probably the most influential naturalist account of health and disorder. According to Boorse, to be healthy is to function normally, and a part of the organism's phenotype is considered to function normally if its contribution to the organism's biological goals of survival and reproduction is statistically typical. Typicality here is defined as typical with regard to a particular reference class, such as species, sex, or age. For example, it is statistically typical for women to have a stable and exclusive sexual preference for men, and this preference contributes to the reproductive fitness of women. However, it is not statistically typical, according to Boorse, for men to have a stable and exclusive sexual preference for other men, and the reproductive success of (exclusively) homosexual individuals is lower than that of heterosexual individuals. Thus, for Boorse, (exclusive) homosexuality is a pathological deviation from the biological norm because "it can hardly be denied that one normal function of sexual desire is to promote reproduction" (Boorse 1975, 63). Although Boorse defends a naturalist account of disorder, his account of the related concept of illness invokes values as well. In his view, an illness is a disorder that is incapacitating or undesirable, and because homosexuality is not always incapacitating or undesirable for the individual, it is not—or at least not necessarily—an illness.

In addition to the naturalist and normativist positions in the philosophical debate about disorder, there are a number of alternative positions, including hybridism. Hybridists argue that disorder judgments depend on both scientific and value judgments about a condition. The philosopher Jerome Wakefield popularized hybridism in the early 1990s, with his so-called harmful dysfunction analysis of the concept of disorder.[9] Wakefield argues that a condition is a disorder if and only if it is both harmful and dysfunctional. He agrees with Boorse that whether a condition is dysfunctional is a scientific fact. Wakefield's definition of dysfunction, however, differs from Boorse's. According to Wakefield, a dysfunctional trait, such as a mechanism, organ, or desire, is a trait that systematically fails to perform its naturally selected function. Moreover, it requires a value judgment to determine whether it is harmful (Wakefield 1992).

It is not straightforward to apply Wakefield's analysis to the case of homosexuality. Wakefield's evolutionary account of function and dysfunction seems to entail that homosexuality *is* a dysfunction. First, sex has evolved at least in part for reproduction. Second, homosexual activities do not result in reproduction. Third, some homosexuals tend to be less motivated to have reproductive sex. It is much more difficult, however, to determine whether homosexuality is harmful. Initially, Wakefield was not particularly clear about which values and what kind of harm were relevant for disorders. In his early work, he suggested that the value judgment should be about social or cultural values, without further specification. Still, it would be quite natural to conclude that culturally shared value judgments about the harmfulness of a condition would suffice to make that condition a disorder (in that culture), as long as the condition was also dysfunctional (De Block and Sholl 2020).

More recently, Wakefield has admitted that his initial analysis indeed suggested that homosexuality is harmful in homonegative societies. Coupled with his belief that exclusive homosexuality is an evolutionary dysfunction, this suggestion leads to the conclusion that homosexuality is a disorder—a bullet Wakefield does not want to bite. He argues instead that homonegativity is misguided and should be replaced with more accurate value judgments: "The social judgment that a condition is harmful may be based on misguided social values, and deeper judgments about what serves justice in the long run can override superficial harm judgments and thus negate disorder attribution" (Wakefield 2013, 34). Obvious examples of misguided value judgments include oppressive claims like "homosexuality is disgusting." However, some examples are less obviously oppressive, such as the claim that homosexuality is harmful because male homosexual couples cannot share biological parenthood. In Wakefield's view, people overvalue having biological children, and it is therefore misguided to think that homosexuality is harmful because two men cannot impregnate each other. What, then, are the relevant "deeper" value judgments to which he refers? According to Wakefield, the APA provided one such judgment in its dealings with homosexuality: "Psychiatrists avoided the incendiary issue of whether homosexuality is caused by a dysfunction and instead overrode the traditional reproductive-harm value claim, arguing that what really matters from a values perspective is capacity for loving human relationships" (Wakefield 2013, 34). In short, homosexuality is not a disorder because it is not harmful, and it is not harmful because it does not interfere with the valuable capacity to love another human being.

Wakefield's argument appears promising, but on closer inspection, it soon runs into trouble. First, Wakefield seems to suggest that the APA removed homosexuality from their manual because they were normativists or hybridists with a true understanding of values. As we have seen, however, the APA has usually presented the prompt for its decision as the adoption of a naturalist account of mental disorder. Second, even though homosexual people do not differ from heterosexual people in their capacity for loving human relationships, it is unclear how that would undo the purported harm of not having biological children with the person one loves. Isn't it still possible that people feel this is a downside of their sexual orientation? And if so, why then wouldn't this suffice to categorize homosexuality as harmful? Third, Wakefield states that a condition can only be deemed harmful when true values are compromised. This leaves room, however, for disagreements among philosophers, psychiatrists, and laypeople about what constitutes a real value and which of those values are relevant here. Moreover, not all philosophers subscribe to the view that there are objectively true values. Indeed, many philosophers reject this view, which is termed moral realism.[10]

Importantly, it is the value criterion and not the scientific criterion that saves homosexuality from being a disorder in Wakefield's hybrid account. Wakefield suggests—even though he never really argues for it—that some forms of homosexuality are dysfunctional, at least according to his evolutionary account of function. In the previous chapter, we noted how studies consistently show that contemporary Western male homosexuals have fewer children than contemporary Western male heterosexuals (see also De Block and Adriaens 2016). So, in a sense, Wakefield remains close to a long naturalist tradition that considers homosexuality a disorder.

Of course, very few contemporary philosophers would be willing to conclude explicitly that homosexuality is a disorder, and for most this conclusion is disagreeable. Wakefield, for example, considers it a successful counterexample to an underdeveloped version of his harmful dysfunction analysis, and even Boorse seems to conclude only reluctantly that homosexuality is a disorder. (Levin is the proverbial exception proving the rule.) In the next sections, we explore why most philosophers and psychiatrists today prefer not to pathologize homosexuality, why this reluctance is a good thing, and how it should be taken into account in any analysis of the concept of disorder.

THE PERILS OF PATHOLOGIZING HOMOSEXUALITY

We mentioned before that there have historically been three negative attitudes toward sexual deviance, including both human and animal homosexuality: criminalizing, moralizing, and pathologizing. We also touched on the often complicated structural and historical relations among these three attitudes. Take, for instance, the relation between moralizing and pathologizing. In recent decades, it has become morally problematic to deny that some conditions, such as depression or fibromyalgia, are disorders. Many mental health awareness campaigns emphasize that mental disorders are real disorders that should be treated similar to somatic disorders like cancer and pneumonia. This classification is in part because patients cannot take full responsibility for their symptoms, such as thoughts, behaviors, and moods. Conversely, it has become morally problematic to affirm that homosexuality is a disorder, even though many of today's gay activists and mental health awareness campaigners share the same ambitions: reducing stigma, culpability, and blame. Why is it that pathologizing sometimes breaks with moralizing and at other times they are perfectly in line?

One of the best illustrations of the ambiguous relation between moralizing and pathologizing can be found in psychiatry's dealings with homosexuality in the past 150 years. In a way, pathologizing homosexuality simply continued an age-old tradition of judging and controlling people's sexuality—a tradition initiated and supported by both ordinary and religious morality as well as the criminal justice system. It is perhaps too easy, however, to lump these institutions together, as the types of judgment and control can make a huge difference for the individual. In countries across the world, homosexuals continue to suffer because authorities and individuals ostracize, persecute, and even execute them. Compared to this oppression, the experiences of victims of the disorder view, who were treated medically or psychotherapeutically for their sexual orientation, seem much less extreme. Today, we may rightfully condemn all three negative attitudes alike, but for nineteenth-century homosexuals, being considered ill rather than abject or criminal often came as a relief. That relief was a partly unintended but nonetheless welcomed effect of the disorder view of early psychiatrists, like Krafft-Ebing. In *Stepchildren of Nature*, Harry Oosterhuis (2000) documents how many homosexuals welcomed the disorder view of homosexuality and notes that Krafft-Ebing was proud of how he contributed to the well-being of homosexual people. Many of them wrote letters to Krafft-Ebing, thanking him for the work he had done and for showing them (and the whole world) that they were not alone. They wrote that they finally felt understood. In one letter, a homosexual physician noted: "I lost all control, and thought of myself only as a monster before which I myself shuddered. Then your work gave me courage again, and I was determined to get to the bottom of the matter, examine my past life, and let the results be what they might be.... After reading your work I hope that ... I may still count myself among human beings who do not merely deserve to be despised" (quoted in Oosterhuis 2000, 11).

One may ask, however, if pathologizing homosexuality was indeed a step forward. Sometimes pathologizing is little more than a subtle form of moralizing, and this subtlety can be more dangerous, as underhanded forms of control are harder to expose (Löfström 1997).[11] Unfortunately, it took a long time—roughly a century—before the medical community started to realize that pathologizing homosexuality was a moral problem.

Why is it bad to pathologize homosexuality, given that it often saved homosexuals from the criminal justice system and reduced their guilt and shame? The answer may be simple. All else being equal, it is usually better not to have a disorder. As Rachel Cooper (2002, 271) puts it succinctly, a disorder is generally "a bad thing to have."[12] Pathologizing creates pressure to conform to particular norms, so calling homosexuality a disorder suggests that it is less valuable or desirable than heterosexuality and that people should try to be heterosexual. Similarly, saying that depression is a disorder amounts to saying that it is generally not good to have depression; it makes sense then to try to "shake it off" or to treat its most undesirable symptoms with, for example, antidepressants or psychotherapy.

If homosexuality is a disorder, many will think of it as "a bad thing to have," which means it is not something to be proud of. That in itself may have a pathogenic effect on homosexuals, as Marmor and colleagues argued in the 1970s. They knew that internalized societal norms might result in psychological conflicts and serious psychiatric symptoms. Homonegative societal attitudes and beliefs can cause homosexuals to feel ashamed and disgusted about their sexual orientation, and this self-loathing may even eventually lead to a true mental disorder. Labeling a condition as a disorder can create unpleasant and stigmatizing effects for individuals with that condition.

Most of these considerations about the morality of pathologizing are based on the assumption that homosexuality is not in fact a disorder, in which case the disorder view is wrong on two counts as both factually inaccurate and morally problematic. Earlier, however, we noted that some naturalist philosophers subscribed and even continue to subscribe to the disorder view, even though they seem to be aware that such pathologizing may negatively affect homosexuals. In a way, they are similar to the prototypical philosopher who follows the argument wherever it may lead. The question then becomes if the moral and the factual can and should be kept apart when defining disorders.

In the next section, we argue that the great harm created by categorizing homosexuality as a disorder should guide our search for a philosophically sound definition of disorder. We can only do this, however, if we rethink the very foundations of conceptual analysis.

CONCEPTUAL ANALYSIS AND BEYOND

Both Boorse and Wakefield are naturalists in that they believe science can help us distinguish between mental health and mental disorder. Yet Boorse's analysis and some interpretations of Wakefield's analysis lead to conclusions that are at odds with both current medical practice and our intuitions. In Western Europe and North America, most people agree with psychiatrists that homosexuality is not a disorder. To see why this dispute is a problem for naturalist analyses of the concept of disorder, we must consider why philosophers engage in such conceptual analysis in the first place.

One of the purposes of conceptual analysis is to break a larger concept down into more simple concepts. These simpler concepts provide expressions of the necessary and sufficient conditions for falling within the original concept. A successful analysis explains both how we tend to use the concept—the descriptive project—and how we should use it—the normative project. By indicating how to properly use the concept of disorder, the normative project helps us, among other things, decide whether particular borderline cases are disorders, such as obesity or attention-deficit hyperactivity disorder. Of course, we do not always need conceptual analysis to clarify which conditions are disorders; we can readily assess that lung cancer and schizophrenia are disorders and that having green eyes and dark hair is normal. If anything, these prototypical conditions are the raw material of conceptual analysis.[13] As Wakefield (1992, 233) explains: "Proposed accounts of a concept are tested against relatively uncontroversial and widely shared judgments about what does and does not fall under a concept." Conceptual analysis is therefore useful not because it can lead to a radical revision of our most basic intuitions about disorder and health but because it can help us decide whether controversial cases are genuine disorders. The main purpose of conceptual analysis is to draw lines between health and disease in the most challenging cases (Ruse 2012).

One important question here is what counts as a prototypical disorder or, in Boorse's vocabulary, as "a paradigm object of medical concern" (Boorse 1977, 545). How do we determine whether a condition is a prototype or a borderline case? Wakefield and Boorse refer to shared judgments or intuitions, but they stay silent on how widely shared these intuitions should be. It is also not clear whether some intuitions—for example, the intuitions of medically trained people—should count more than others. These are important questions, as relevant intuitions or judgments do differ, even among specialists (De Block and Sholl 2020). Disagreements can be fundamental, with some considering a condition a prototype, and hence as an illustration of the truth or falsehood of a proposed analysis, while others may consider it a borderline case. If it is a borderline case, then the results of the conceptual analysis will be used as tools to determine whether the condition should be considered as healthy or disordered.

Homosexuality is a case in point. Wakefield considers it a straightforwardly healthy condition and treats it as a potential counterexample that undermines rival analyses as well as some interpretations of his own harmful dysfunction analysis. As we noted earlier, the very possibility that some might interpret the harm component of Wakefield's analysis to pathologize homosexuality has urged him, in recent years, to reformulate and revise that component (Wakefield 2013; Wakefield and Conrad 2020). Boorse, by contrast, considered homosexuality a borderline case that, when confronted with his biostatistical account, turns out to be a disorder. Even though he has always been slightly uncomfortable with this conclusion, Boorse has defended it because

> recent debates over homosexuality and other disputable diagnoses usually ignore at least one important issue. Besides asking whether, say, homosexuality is a disease, one should also ask what difference it makes if it is. I have suggested that biological normality is an instrumental rather than an intrinsic good. We always have the right to ask, of normality, what is in it for us that we already desire. If it were possible, then, to maximize intrinsic goods such as happiness, for ourselves and others, with a psyche full of deviant desires and unnatural acts, it is hard to see what practical significance the theoretical judgment of unhealthiness would have. (Boorse 1975, 63)

In Boorse's view, disorder judgments should be clearly distinguished from both moral and clinical judgments. Whereas the latter usually require action, a disorder judgment in itself does not. Boorse believes that having a disorder is not always a bad thing.

A related question is whether moral considerations should play a role in the analysis of the concept of disorder. To be clear, this is not a question about a specific conceptual analysis of disease but a question about what conceptual analysis should be. In other words, the question is not about whether moral concerns are central to our intuitions about disorder judgments, as normativists claim and naturalists deny. The question here is rather whether we *should* consider moral concerns when we ask whether a condition is healthy or disordered. Are there good moral reasons for considering homosexuality as a prototypically healthy condition, even when our intuitions about it may still be vague or conflicting? Or should we side with Boorse, who suggests that only our intuitions (and not our values) should determine whether something is a clear-cut or a borderline case?

In our view, there is much to say in favor of a value-based approach to prototypical cases. In fact, this kind of approach has become increasingly popular within philosophy. For instance, Eric Schwitzgebel (forthcoming) has argued for what he calls a "pragmatic" approach. He explains this approach as follows: "If there's more than one way to build a ... classificational scheme, you have some options. You can consider, do you want to classify the thing in question as an A? Would there be some advantage in thinking of category 'A' so that it sweeps in the case? Or is it better to think of 'A' in a way that excludes the case or leaves it intermediate? Such decisions can reflect, and often do at least implicitly reflect, our interests and values."

The values Schwitzgebel mentions certainly include moral values. In the context of the analysis of the concept of disorder, the pragmatic approach implies that such analysis can only be successful if it (a) is in line with our uncontroversial intuitions about healthy and disordered conditions and (b) classifies intuitively vague or controversial cases in a way that fits our (moral) interests and values. The first requirement is shared with traditional conceptual analysis, whereas the second requirement sets the pragmatic approach apart.

In the next section, we argue that there is a clear advantage in thinking of the categories of health and disorder in such a way that homosexuality is a healthy condition, even though homosexuality may appear to some people as a borderline case. In other words, even though intuitions may disagree on whether homosexuality is healthy, there are good reasons to see homosexuality as a prototypically healthy variant of human sexuality. Again, if we adopt such a pragmatic approach, the goal can never lie in deciding whether homosexuality is a disorder. Rather the goal is to develop a conceptual analysis of disorder that takes the healthy status of homosexuality as a fundamental and morally motivated constraint.

THE SHORT PHILOSOPHICAL ROUTE FROM ELIMINATIVISM TO PRAGMATISM

Importantly, Schwitzgebel's pragmatic approach (like Sally Haslanger's [2000] closely related "ameliorative project") does not exclusively aim at analyzing our ordinary use of concepts. This approach also reflects on what work we want these concepts to do. Because it matters to people whether we classify conditions as disorders, it seems legitimate to take that "mattering" into account when we try to prescribe how these labels should be used. Of course, that requires a good understanding of how and why these labels matter. Are these labels significant primarily for scientific or for moral or existential purposes?

The philosopher Germund Hesslow (1993) argued persuasively that the distinction between health and disorder has very little theoretical or clinical importance. If it did, Hesslow claimed, medical professionals would have joined the philosophical debate long ago, looking for guidance in their medical research or practice. Generally, however, there is no clear connection between disorder judgments and therapeutic decisions, nor between disorder judgments and biomedical theories on the etiology and development of disorders. Benign tumors are often listed as disorders, for example, but some of them are left untreated. Conversely, plastic surgeons provide hair transplants for bald men, even though baldness is rarely listed as a disorder.

More recently, philosopher Marc Ereshefsky (2009) took Hesslow's position to its eliminativist extreme, arguing that if the concepts of health and disorder really do not matter, then we should stop using them. Of course, even in an eliminativist world, there is still a lot that psychiatry can and should do. According to Ereshefsky, psychiatry should refocus on describing and explaining conditions. Additionally, it should answer questions as to how these conditions are valued and whether it is wise to treat them. Instead of asking whether a condition is a disorder, according to Ereshefsky, medicine and psychiatry should attempt to describe and explain the condition properly and to chart the relevant value judgments about this condition. One can chart these judgments by asking a number of questions: Is the individual at ease with the condition? How does society perceive the condition? How have these individual and societal judgments come about? Still, even if the individual thinks that the condition is undesirable, psychiatrists need not necessarily intervene. They may judge, for example, that an intervention will have little or no effect, that it will actually do more harm than good, or, more importantly, that the individual's social environment is the cause of his suffering. Many of the psychiatrists discussed in this chapter realized that the suffering of homosexuals often finds its origin in other people's negative judgments. In such cases, psychiatrists can be of help by, for example, advising their patients to look for a different environment or helping them guard against negative judgments.

However attractive and convincing the eliminativist position may be, we think that a pragmatic approach to health and disease is still preferable. The concepts of health and disorder may not do much work in clinical research and practice, but these concepts do a lot of work in various other contexts, including our everyday life. Therefore, it seems unlikely that they will soon disappear altogether, and thus they will continue to impact individuals and communities. For example, cognitive scientists and ethicists have documented clear labeling effects showing that both pathologizing and depathologizing can reduce or increase stigma, depending on the context and the condition (Mehta and Farina 1997; Rüsch, Angermeyer, and Corrigan 2005; Goldberg 2014). If we want our concepts to do good moral work and reduce rather than increase stigma, it seems reasonable to try to curb the concepts to better fit these moral concerns. This is certainly no plea for a complete revision of our concepts or intuitions; it is rather a call to let moral concerns play a role in determining which outcomes of our conceptual analyses should be avoided. By doing so, preferred avoidable outcomes are not just those that go against our most shared and strongest intuitions about the correct use of the concept (as in traditional conceptual analysis) but also outcomes that are morally undesirable.

Because moral effects are not always straightforward and often involve many stakeholders, pragmatism is hard to implement. Moral considerations will often be as messy and conflicting as our intuitions about health and disorder. Hence, implementing this proposal would not necessarily make it easy to define disorder, but it would shift at least part of the discussion to more pertinent issues than those currently at play in the philosophical debate. Much of the discussion on what constitutes a disordered or healthy condition would then center on the moral values that are at stake in calling controversial conditions healthy or disordered and on who is affected by the relevant judgment. Debates over the medicalization of low mood, for example, would mostly focus on how to compare the moral costs and benefits of categorizing low mood as either a disorder or a social problem (Stegenga 2018). Such debates would be about how we should perform this moral calculus and maybe even about whether such calculus is possible in a particular case.[14]

Importantly, however, there would not be much to discuss in the case of homosexuality: most psychiatrists and laypeople agree that calling homosexuality a disorder will—all things considered—cause more harm than labeling it as a healthy, normal condition. We admit that the situation may be different under regimes that persecute homosexuals or in cultures in which they are still ostracized.[15] The cultural relativity of the disorder category does not pose deep philosophical problems for our proposal. Different contexts often result in a different moral calculus, and disorder judgments are environmentally relative, as even some defenders of traditional conceptual analyses have acknowledged (Matthewson and Griffiths 2017).

Before we conclude this section, we want to emphasize that a pragmatic approach to the concept of disorder is not to be confused with normativism or eliminativism. Traditionally, normativist accounts of disorder state that moral and/or existential norms best capture the distinction between health and disorder, while naturalistic accounts claim that only scientific facts are relevant for that distinction. The pragmatic approach does not decide between these two types of account; it only redefines how we should assess them. According to the traditional approach to conceptual analysis, an analysis is successful if it captures our intuitions about uncontroversial disorder and health ascriptions. If it succeeds in doing so, it can then be safely applied to the vague and controversial cases. The debate between naturalists and normativists currently only focuses on whose account best captures these intuitions. Within a pragmatic project, however, the assessment involves more than just intuitions; it also involves values. A pragmatic analysis—be it a normative or a naturalist analysis—can only be successful both if it is in line with our uncontroversial intuitions about healthy and disordered conditions and if it classifies intuitively vague or controversial conditions in a morally desirable way. The pragmatic approach agrees with traditional conceptual analysis that it would not be reasonable to classify green eyes as a disorder or pneumonia as a healthy condition, but unlike the traditional approach, the pragmatic approach demands not that morally charged borderline cases are decided by the analysis but rather that they *steer* the analysis.

A pragmatic approach also differs from eliminativism. Eliminativism proposes to do away with conceptual analysis; after all, analyzing concepts that we should eliminate from our language and thinking is a waste of time. The pragmatic account thinks that it is almost impossible to eliminate these terms and concepts, and it is therefore valuable to analyze what they do and should mean.

CONCLUSION

The psychiatric history of homosexuality provides an interesting illustration of how medical intuitions and the values that shape these intuitions have changed over time.[16] These historical changes make it very difficult to give an ahistorical evaluation of how nineteenth-century and twentieth-century psychiatry dealt with sexuality in general and homosexuality in particular. Inevitably, the context of a society determines its values and intuitions. Still, we can assess what it meant for people to have their sexual orientation labeled a disorder. For nineteenth-century homosexuals, the psychiatric attention probably had its benefits, although it is hard to weigh these against the many drawbacks. After World War I, these benefits disappeared almost entirely, and only the downsides remained or became even greater. The oppression of homosexuals under the Nazi and Soviet regimes, for example, was not lessened by the fact that these regimes considered homosexuality a disorder. In fact, psychiatric labels were seen as a reason to treat homosexuals as subhuman (Greenberg 1990) in both Nazi Germany and the Soviet Union.

In the early 2000s, Glenn Smith and his colleagues interviewed twenty-nine homosexuals who had undergone medical treatment throughout their lives because of their sexual orientation. Their testimonials are heartbreaking (G. Smith, Bartlett, and King 2004). Often, those who are ostracized and humiliated seek the company of fellow sufferers to share their experiences and seek compassion as well as to try to undo the injustices that they suffered. This community formation can then form the basis for the further development of a common identity. In that sense, the pathologizing of homosexuality probably seeded its own demise. Indirectly, it also spurred a wider scrutinizing of psychiatry and the nature of its disorder judgments. To this day, philosophical debates on the definitions of disorder and health are often explicitly inspired by how psychiatry has dealt with homosexuality. We hope that this inspiration will eventually lead to definitions of health and disorder that unambiguously place homosexuality on the healthy side of the divide.

Epilogue

Gaydars and the Dangers of Research on Sexual Orientation

One of the conclusions to be drawn from this book is that, for a very long time, homonegative attitudes and beliefs have pervaded large parts of the gay sciences. Findings from these sciences were regularly used to legitimize the stigmatization and persecution of homosexuals. Our chapter on animal homosexuality, for example, clearly testifies to rampant homonegativity in nineteenth-century entomology, while critics branded Franz Kallmann's work on the genetics of homosexuality as dangerous and stigmatizing pseudoscience (Lewontin, Rose, and Kamin 1984). Some recent commentators, including Joan Roughgarden (2017), have even argued that homonegativity still haunts the biological gay sciences today. In addition, the previous chapter shows that the situation has at times been even worse in psychiatry.

Then again, since the end of the nineteenth century, both biology and other gay sciences have been instrumental in the emancipation of homosexuals, with scientists from various backgrounds presenting rather robust data to combat homonegative attitudes and beliefs. Examples of such beliefs include the idea that homosexuals seduce or recruit others into homosexuality and the idea that a homosexual orientation can (and needs to) be changed through therapy (J. Bailey et al. 2016).

There are indeed all sorts of interesting connections between views on the nature and origin of homosexuality, on the one hand, and emotional and moral attitudes toward homosexuality, on the other. For medieval theologians, for example, to understand the nature of homosexuality was to condemn it as unnatural. More recently, the relation between facts and moral attitudes has loosened, and many now believe that facts about the nature and origin of homosexuality are largely irrelevant vis-à-vis its moral value. It is surprising, then, that in recent years some philosophers and queer theorists have branded scientific research into the causes of homosexuality as intrinsically homonegative, with some even arguing such research should not be conducted at all.

In this epilogue, we explore how value neutral the gay sciences should be. First, we discuss the fusion of the natural and the moral in natural law thinking, as well as its implications for the morality of homosexuality. Second, we argue that elements of natural law thinking are still part of our intuitive moral psychology of sexual orientation. Third, we ask whether we should trust or distrust our moral intuitions in this particular context. We conclude with an assessment of the claim that some research on homosexuality should not be done.

AGAINST NATURE

Western history has previously known periods of relative tolerance toward at least some kinds of homosexuality. Today's tolerance, however, is unprecedented. Many believe its increase relates to the process of secularization, and it is indeed an understatement to say that Christianity has not always been kind to homosexuals. Leviticus famously denounces homosexual activities as "detestable" (Lev. 18:22) and considers them a capital offense: "If a man sleeps with a man as with a woman, they have both committed a detestable thing. They must be put to death; their blood is on their own hands" (Lev. 20:13).

However, these Old Testament pronouncements might be more ambiguous than they seem. Some biblical scholars have argued, for example, that these laws relate primarily to sex with boys or to anal sex, rather than to sex between adult men (Olyan 1994; see also Douglas 1999). Moreover, it is unclear how Old Testament instructions should apply to Christians, since a large number of Leviticus's prohibitions were not seen as binding for them. Others have added that, despite Paul's fierce condemnation, the New Testament is more tolerant, with some interpreting Jesus's silence on homosexuality as a sign that he was not very concerned about it (Boswell 1980). Nevertheless, there is a consensus that both the Jewish tradition and early Christianity were far less tolerant than the ancient Greeks. The Romans were already less permissive, referring to homosexuality as the "Greek affliction"—a foreign habit to be condemned and punished, though relatively mildly. This mildness slowly disappeared under the influence of the Stoics, whose aversion eventually culminated in the openly homonegative views and laws of the first Christian theologians and emperors (Kuefler 2007).

Early Christianity considered homosexual behavior a sin against nature. Augustine, for example, described homosexuality as a vice because it is based on a desire for unnatural excess (Anagnostou-Laoutides 2015). The condemnation of homosexuality escalated throughout the twelfth and thirteenth centuries, spurred on by developments in theological and philosophical accounts of so-called natural law, including the work of Thomas Aquinas (Jordan 1997; Boswell 1980). According to Aquinas's natural law theory, moral norms are derived directly from nature. The nature of the world and our human nature determine what is good and what is bad. Human genitals, for example, were created to generate new life, and that generative purpose determines how they should be used. According to Aquinas, only generative sexual acts, that is, acts that may result in offspring, can be morally acceptable. Clearly, homosexual activities are not generative, and therefore Aquinas considered them immoral. To be clear, Aquinas did not believe that all generative acts were morally permissible, nor did he think that all sexual acts should result in offspring. First, some generative acts, such as adultery and rape, are obviously immoral. They are immoral not because they go against the nature of sexuality but because they are harmful to fellow human beings. Rape harms the victim, while adultery harms the child who may be born out of wedlock. Second, some sexual activities that do not result in offspring are still morally acceptable, as long as they happen between consenting adults within marriage. In Aquinas's view, sexuality is generative because of its potential, rather than because of its actual effects (pregnancy and offspring). We are morally obligated to go in the direction of the end indicated by the nature of sexuality, but we are not morally obligated to achieve that end.

Aquinas's version of natural law theory is rather audacious, as it derives moral values directly from biological functions. In his view, human male homosexuality is immoral because it does not respect the natural, generative function of sexuality. Today, few philosophers side with Aquinas, and their reservations often originate instead in David Hume's famous critique. According to Hume, what we *ought* to do (moral norms) cannot be logically derived from what *is* (natural facts). It may be in our human nature to be selfish, for example, but that does not imply that we should all strive for selfishness. Likewise, even though some homosexual activities may not seem very useful from a biological perspective (see, however, chapter 3), it does not follow that these activities are morally reprehensible. In Hume's view, biological functions and moral values belong to two fundamentally different categories.[1]

Even though natural law theory is no longer popular among contemporary philosophers, it continues to be important in understanding the moral beliefs of many nonphilosophers. In fact, it is so persistent that the Australian philosopher Paul Griffiths (2002) once claimed that humans are natural law thinkers *by nature*. In his view, even people who grow up outside the purview of Greek philosophy or Christianity will almost inevitably have a moral preference for what they consider natural (see also S. Scott and Rozin 2020). This natural law logic often reveals itself in laypeople's opinions about the moral value of homosexuality. In our general introduction, we mentioned the catalogue of a successful recent exhibition on animal homosexuality, in which the author expressed his hope that the existence of animal homosexuality would help "reject the all too well-known argument that homosexual behavior is a crime against nature" (Søli 2009). In the next section, we map out the relationships between people's moral attitudes toward homosexuality and their views on its nature and etiology.

OUR TOLERANT NATURES

Psychological research on homonegativity has been inspired by earlier psychological research on racism. Social psychologists have discovered that the same factors that correlate with racism, such as an authoritarian personality or conservativism, also correlate with homonegativity (Altemeyer 1996; Herek 1994; Sakalli 2002). For example, men are on average more racist and more homonegative than women, as are people with distinctly authoritarian personalities. Still, there are interesting differences between the correlates of racism and homonegativity (Cottrell and Neuberg 2005). The most important difference is in the way they correlate with biological explanations, with the correlation mostly positive in the case of racism and mostly negative in the case of homonegativity. In this section, we attempt to unravel these correlations.

In his now-classic *The Nature of Prejudice*, the psychologist Gordon Allport (1954) revealed that essentializing or naturalizing social categories is a major source of prejudice. Racism rises with the belief that there are deep biological differences among populations defined as races. Today, Allport's work is echoed in research on psychological essentialism. Psychologist Douglas Medin and philosopher Andrew Ortony first defined "psychological essentialism" by comparing it with "metaphysical essentialism," which they described as the view that "an object has an essence by virtue of which it is that object and not some other object" (Medin and Ortony 1989, 183). The reader may recall that we previously accused some evolutionary biologists of essentializing homosexuality in this way. Psychological essentialism, by contrast, "is not the view that things have essences, but rather the view that people's *representations* of things might reflect such a belief (erroneous as it may be)" (183).

According to Medin and Ortony, our intuitive classification of the world is based on perceived similarities among things, as well as on theoretical beliefs about the underlying characteristics thought to cause these similarities. These beliefs are essentialist beliefs in that they attribute some kind of inherent essence to the classified objects. Animated and nonanimated entities are what they are not because of their easily observable properties but because of an invisible and core essence, like those that philosophers have discussed since Plato.

Psychologist Susan Gelman has shown that even very young children judge that the identity of an object is usually not determined by visible and superficial properties (S. Gelman and Wellman 1991; S. Gelman 2004). A beetle that looks like a leaf is seen more as a beetle than as a leaf, even though the superficial similarity with a leaf is in fact far more salient than the similarity with a beetle. A beetle is a beetle not because of its observable qualities but because of its beetle essence, the possession of which is a necessary and sufficient condition to be considered a member of the beetle category.

Building on many experiments like these, Gelman (2003, 6) argued that essentialization is "deeply rooted in our conceptual system and reflected in very different cultural contexts at a very young age." From an early age, people from different cultures tend to classify the natural and social world into essentialist categories. Animals and fruits and even ethnic groups and sexual orientations are identified and distinguished from one another based on an alleged underlying essence that defines category membership. If we think that skin color is caused by some deep, invisible essence shared by all people of a particular race, then it makes sense to use skin color to divide people into different races, even if we do not think that skin color alone (or any other superficial property purportedly caused by this essence) defines race. Likewise, if we think that a preference for pink is caused by the underlying essence of homosexuality, then it makes sense to use color preferences to classify people's sexual orientations.[2]

Recent research into psychological essentialism and prejudice confirms Allport's idea that essentialist beliefs about a social group often come with negative attitudes toward people in that group (Jayaratne et al. 2006; Keller 2005; M. Williams and Ebenhardt 2008). According to psychologist Vincent Yzerbyt (Yzerbyt, Rocher, and Schadron 1997), essentialism prompts stereotyping and prejudice because it legitimizes the status quo by presenting it as natural, and what is natural is usually very hard to change. If you have a biological essence, and that essence makes you what you are by determining your behavior and attitudes, it is almost unthinkable that you would act against it. Yzerbyt argues that essentialism plays an important role in both sexism and racism. After all, if you believe that women are naturally less ambitious, for example, then you will probably oppose affirmative action for women, since affirmative action would force women to go against their true and essential female preferences. Similarly, if you believe that people who belong to other races are different by nature, then you will likely explain racial differences in attitudes and aptitudes as an inevitable consequence of immutable racial essences. According to that mindset, a higher crime rate among Black people, for example, is not so much the result of segregation, discrimination, or other sociocultural factors but rather the natural expression of some inner natural quality ("It's in their blood," "It's in their genes," etc.). In brief, essentialism about human races increases racism.

Remarkably, that correlation is not as straightforward in the context of sexual orientation, as research indicates that psychological essentialism about homosexuality tends to be associated with *reduced* homonegativity (Hegarty and Pratto 2001; Morton, Hornsey, and Postmes 2009). There is a positive correlation, for example, between essentializing homosexuality and tolerating same-sex marriage and same-sex adoption. In other contexts, however, psychological essentialism seems to increase homonegativity (De Cecco and Elia 1993).

To make sense of the complex relation between homonegativity and psychological essentialism, psychologists Nick Haslam and Sheri Levy empirically distinguished among three different clusters of essentialist beliefs about homosexuality (Haslam and Levy 2006). The first cluster, similar to the fixity interpretation of innateness discussed in chapter 1, includes the beliefs that homosexuality is determined in early childhood; that it is immutable, at least at the level of the individual; and that it has a biological basis. The second cluster includes the beliefs that homosexuality is fundamentally different from heterosexuality and that this difference relates to their essences. The third cluster includes the beliefs that homosexuality has always existed and that it occurs in all cultures. Beliefs from the first and third clusters negatively correlate with homonegativity: the more intense these beliefs, the *less* homonegativity one expresses. However, beliefs from the second cluster have a positive correlation with homonegativity: the more intense these beliefs, the *more* homonegativity one expresses.

Haslam and Levy explain their findings as follows. Beliefs from the first cluster lead to less homonegativity because we do not tend to hold others responsible for the properties we consider to be biologically determined.[3] A similar logic applies to the third cluster, which emphasizes immutability, be it now at the population level. The second cluster, however, emphasizes the differences between heterosexuals and homosexuals. This type of essentialism increases homonegativity because it assumes homosexuals belong to a fundamentally different group of people—a view that often prompts out-group hate. Haslam and Levy (2006, 483) suggest that "heterosexual men may view gay men as categorically different to distinguish themselves sharply from a despised identity" and that "some essentialist beliefs serve a 'boundary reinforcement' function, sharpening a distinction so as to safeguard the person's identity" (see also D. L. Smith 2014).

At first glance, Haslam and Levy's explanation makes sense. We should not forget, however, that their research is purely correlational, so it only examines to what extent the two variables (essentialist beliefs and homonegativity) go together, not to what extent they influence each other, let alone in which direction the influence goes. In other words, Haslam and Levy's explanation hints at causal relationships that are hard to substantiate with purely correlational research.

From other experimental research on this topic, we know that the causal arrow can go in many directions. There is some evidence that the experimental manipulation of essentialist beliefs can indeed bring about changes in attitudes (Ramirez 2007). For example, Oldham and Kasser (1999) found that reading a vignette that focuses on the genetic and neuroendocrine causes of homosexuality led to less homonegativity in students, and this effect, though rather weak, lasted for more than a few days. Other studies did not find such causal correlation (Piskur and Degelman 1992; Pratarelli and Donaldson 1997). To complicate matters, still other studies have revealed that moral attitudes about homosexuality often seem to influence which beliefs about its nature we adopt. People rationalize their preexisting attitudes by endorsing certain statements about the nature and origin of homosexuality (Boysen and Vogel 2007; Hegarty and Golden 2008).

Should we be worried, then, about the prospects of combating homonegativity with science? Given the messy relation between psychological essentialism and homonegativity, should we be skeptical about Magnus Hirschfeld's motto that science will lead to justice (*Per scientiam ad justitiam*)? Generally, we believe that informing people about the nature and origins of homosexuality will somewhat increase the acceptance of homosexuality (see also J. Bailey et al. 2016). One should not forget, however, that there is still considerable scientific debate about the nature and origins of homosexuality. It would be rash to assume that the "scientific" statements used in studies on psychological essentialism and homonegativity reflect scientific consensus. Throughout this book, we have argued that part of the variation in sexual orientation is explained by genetic and hormonal variation, but we are still far from a clear and definitive view on homosexuality's etiology. It may well be the case that homonegativity decreases when people are informed that homosexuality is caused by genes and hormones, but their acceptance is not necessarily due to accurate information. The etiology of homosexuality is likely much more complex than what is suggested by the vignettes used in psychological studies. As one prominent biologist has admitted, our current biological knowledge does not rule out the possibility that "all the biological factors . . . only produce a predisposition to become homosexual, and [that] these predispositions can only develop in a specific set of psychosocial contexts that are not yet identified" (Balthazart 2011b, 159).

In addition, even if gay activists and scientists have science on their side, it remains to be seen whether science is responsible for most of the decrease in homonegativity. In so-called "Western, Educated, Industrialized, Rich, and Democratic" societies, to borrow Joseph Henrich's (2020) phrase, the decline of homonegativity may also be an integral part of a well-documented and broad tendency toward individualizing values and away from binding values that revolve around purity, authority, and in-group loyalty (Barnett, Öz, and Marsden 2018; C. Ellis and Stimson 2012; Talhelm 2018). If that is the case, then the increase in tolerance toward homosexuality is best explained by individualism, rather than by specific beliefs about the etiology of homosexuality (Eriksson and Strimling 2015).

It is one thing to be moderately skeptical about Hirschfeld's idea that, at the end of the day, studying the roots of homosexuality will benefit homosexual people. It is quite another thing to argue that such a scientific project is in itself a homonegative endeavor. Still, that is exactly the argument that prominent philosophers and queer theorists have made. In the remainder of this epilogue, we discuss and criticize their views.

IS IGNORANCE BLISS?

Philosopher Philip Kitcher (2001) has repeatedly argued against research into racial differences. According to him, we should not study cognitive and behavioral differences among ethnic groups because such research has historically been harmful to Black people. For similar reasons, philosopher Janet Kourany (2016) has argued against research into cognitive differences between the sexes. Such an argument could also surely be made about much of the research discussed in this book. Prewar genetic research, for example, has been used in various countries, including Nazi Germany, to justify the persecution and killing of tens of thousands of homosexuals, as well as people with disabilities and mental illnesses.

This horrific history of the gay sciences has led some philosophers to criticize research on homosexuality's etiology. Some attempted to debunk specific studies, such as studies on conversion therapies; others have argued against entire research programs, such as heritability studies, or against the usefulness of particular methods, such as animal models (Stein 1999; Fausto-Sterling 1995). Their criticism often alludes to bigotry on the part of the scientists who conduct this sort of research. For example, Vincent

Savolainen and Jason A. Hodgson (2016, 4) claim that homonegativity and heteronormativity are among the mainsprings of contemporary evolutionary theories about homosexuality: "It is likely that the focus on HS [homosexuality] as an evolutionary paradox is largely driven by cultural taboos against the behavior, rather than any real unique biological perplexity." The feminist scholar Valerie Rohy (2012, 114–15) agrees: "Other writers repeat this question [why do 'gay' genes survive natural selection] with a certain relish at the prospect of homosexuals' evolutionary demise and a palpable frustration at its postponement."

If scientists are biased against homosexuality, then we have a good reason to distrust their theories and findings. Consider, for instance, the notorious New Family Structures Study by sociologist Mark Regnerus (2012). In his first paper based on the study, Regnerus concluded that adult children raised by a parent who had a romantic same-sex relationship used more marijuana and were less well-employed than adult children who grew up in families with heterosexual parents. Critics immediately faulted the study's methodology (Cheng and Powell 2015), and many associated Regnerus's questionable methodological choices with his ideological agenda (Musick 2014).

Importantly, however, none of these critics called for a ban on research studying the effect of family structure on a child's well-being. The question remains, then, if there are aspects of homosexuality that we should not study at all. Philosopher David Hull once suggested that the etiology of homosexuality may be one of those aspects. In his view, research on this topic is inevitably guided by the idea that "heterosexuality is the normal state programmed into our genes. It needs no special explanation. Normal genes in a wide variety of normal environments lead most children quite naturally to prefer members of the opposite sex for sexual and emotional partners. Homosexuality, to the contrary, is an abnormal deviation which needs to be explained in terms of some combination of defective genes and/or undesirable environments" (Hull 1998, 390).

Hull's critique echoes historian David Halperin's straightforward claim that "the search for a 'scientific' etiology of sexual orientation is itself a homophobic project" (Halperin 1990, 49). Both Hull and Halperin believe that such research is morally problematic, almost to the extent of being impermissible, because of its heteronormative focus and its homonegative assumptions. In the words of one commentator, these critics' concerns are not primarily with the intentions or motivations of individual researchers. Rather, they are concerned "that the very motivation for seeking the 'origin' of homosexuality has its source within social frameworks that are pervasively homophobic" (Schüklenk et al. 1997, 9). Hull and Halperin never really substantiate their suggestion that etiological research is unethical, but others have supplemented their statements. In this section and the following, we discuss two of their arguments.

A first hypothetical argument for abstaining from research on the causes of homosexuality is analogous to the argument defended by Kitcher (2001) against researching racial differences in cognitive ability. According to this argument, ignorance is bliss when studies are likely to give rise to beliefs that strengthen or stabilize convictions with morally undesirable social consequences. Indeed, Kitcher takes issue with the research project itself, regardless of the findings it would yield. His assessment is that cognitive group differences are probably not due to genetic differences among groups. If research clearly and unequivocally supported Kitcher's assessment, there may not be much of a problem for the underprivileged. However, evidence to the contrary will definitely create additional harm for them, because it will strengthen the attitude that the underprivileged group is less well off not because of discrimination but because they simply lack the abilities required for cognitively demanding social roles. Importantly, according to Kitcher, indecisive evidence will also harm the underprivileged, because he believes our cognitive biases are such that we will interpret indecisive evidence as evidence that supports our gut feelings and biases.[4] Given that science is fallible, and that indecisive or vague findings are hard to avoid in research on cognitive differences, it is better to abstain from this line of research.

Many philosophers have criticized Kitcher's proposal (see, e.g., Talisse and Aikin 2007). Still, even if we accepted it, it is unclear whether it similarly entails that research into the etiology of sexual orientation should be avoided. Does Kitcher's argument also work in this particular context? Surely, some of Kitcher's criteria for impermissible research are met. First, it seems fair to say that homosexuals are still an underprivileged group. In many countries, people continue to be persecuted because of their sexual orientation, and even in many Western countries, homonegativity has not entirely disappeared. Second, negative stereotypes about homosexuals are still prevalent, such as the beliefs that they are dangerous as teachers or that they have narcissistic personalities. Third, as demonstrated throughout this book, etiological research on homosexuality will often yield indecisive results. However, one of Kitcher's criteria does not seem to be met. Earlier, we explained that there is not always a clear causal connection between homonegativity and beliefs about the etiology of sexual orientation. Therefore, we simply do not know whether it would generally benefit homosexuals if people believed that their sexual orientation is biologically controlled. Although findings about the biological roots of homosexuality are often interpreted as indicating that homosexuals should not be held responsible for their sexual orientation (e.g., Swaab 2007), this reasoning itself is characterized by a subtle form of homonegativity. It suggests that we would blame homosexuals if their sexual orientation had been a choice, and that it is obvious that one would prefer to be heterosexual when given a choice.[5]

In short, Kitcher's imperative not to pursue certain lines of research may work for studies on cognitive differences among races but does not necessarily apply to studies on the etiology of homosexuality. In the next section, we consider a second argument for the impermissibility of such studies.

DON'T PUBLISH, AND WE WON'T PERISH

One reason to restrict research is that it produces beliefs that harm the underprivileged. Another reason is that it produces harmful technology. In the past decades, for example, some commentators have argued for constraining biological research on homosexuality because it might lead to screening procedures that would allow prospective parents to select for heterosexual children (Schüklenk et al. 1997).

More recently, academics have argued against research on how machine learning can be used to develop an artificial "gaydar." These critiques were provoked by a recent paper by psychologists Yilun Wang and Michal Kosinski, which was published in the prestigious *Journal of Personality and Social Psychology* (Wang and Kosinski 2018). Wang and Kosinski's research was inspired by the prenatal hormone theory of sexual orientation, which we discussed in several chapters of this book. Crucial support for this theory comes from correlations between sexual orientation and observable characteristics that seem to be determined by prenatal exposure to particular hormones (Berenbaum and Beltz 2011). For example, intrauterine exposure to androgens is thought to influence digit ratio, that is, the length of the index finger divided by the length of the ring finger. On average, male homosexuals tend to have a slightly higher digit ratio than male heterosexuals—a finding many consider evidence for the prenatal hormone theory. Similarly, auditory-evoked potentials, or brainwaves produced in response to auditory stimuli, appear to be stronger in homosexual men than in heterosexual men. One hypothesis holds that this auditory difference is due to intrauterine exposure to androgens (McFadden 2002).

Some believe that the prenatal hormone theory suggests that sexual orientation correlates with facial features. If intrauterine exposure to androgens influences both one's sexual orientation and one's facial features, one can expect that the two will correlate. Wang and Kosinski readily admit that the evidence for such hypotheses is still rather weak, so they call on artificial intelligence to substantiate their research. More particularly, they used artificial neural networks, which are able to learn to detect hard-to-observe patterns. They trained these networks with an enormous amount of data. These data included tens of thousands of pictures found on dating websites and Facebook. Wang and Kosinski established the sexual orientation of the pictured individuals through their profile (for dating websites) or liked pages (for Facebook), such as "Manhunt" and "Gay Times Magazine." According to Wang and Kosinski, their technology could predict a pictured person's sexual orientation based only on their facial features. Importantly, their artificial gaydar did much better than human judges. Given one picture of a person, the algorithm was able to correctly distinguish between heterosexual and homosexual men in 81 percent of cases, whereas human guesses were only correct in 61 percent of cases.

Much like Regnerus's study on homosexual parenting, Wang and Kosinski's gaydar study has been heavily critiqued. First, there were concerns about methodology and data interpretation. Critics argued that the authors' reliance on pictures from dating sites and from openly homosexual social network users was flawed because these pictures likely contain hints about the person's sexual orientation that have nothing to do with facial differences (Fasoli and Maass 2018). A replication study confirmed this idea, demonstrating that the artificial gaydar was still incredibly accurate even when the pictures' faces were blurred (Leuner 2019). Other scholars attempted to show how the artificial gaydar was much less precise than Wang and Kosinski claimed, and that, more generally, it is simply illusionary to think that there is a pure biological measure of sexual orientation (A. Gelman, Mattson, and Simpson 2018).[6]

There were also important moral worries about this study, for instance about privacy and consent. Some commentators contended that Facebook pictures cannot be considered public data: "The LGBTQ community has long had concerns about research, privacy, and consent: we don't treat gay dating profiles or gay bars as public places, and neither should researchers" (Mattson 2017). Another concern related to the potential to abuse Wang and Kosinski's gaydar. Critics argued that this technology could be used by homonegative people and governments to find homosexuals and harm them. Moreover, if the tool was indeed rather inaccurate but perceived as accurate, it could potentially harm heterosexuals as well, if only by generating a large number of false positive results.

Interestingly, Wang and Kosinski anticipated this critique in the conclusion of their paper:

> Some people may wonder if such findings should be made public lest they inspire the very application that we are warning against. We share this

concern. However, as the governments and companies seem to be already deploying face-based classifiers aimed at detecting intimate traits..., there is an urgent need for making policymakers, the general public, and gay communities aware of the risks that they might be facing already. Delaying or abandoning the publication of these findings could deprive individuals of the chance to take preventive measures and policymakers the ability to introduce legislation to protect people. (Wang and Kosinski 2018, 255)

Their main aim was thus to create awareness about how easy it is to develop deep neural networks that are able to detect one's sexual orientation based on nothing but a publicly available picture. In an interview with the *Guardian*, Kosinski added: "I stumbled upon those results, and I was actually close to putting them in a drawer and not publishing—because I had a very good life without this paper being out. But then a colleague asked me if I would be able to look myself in the mirror if, one day, a company or a government deployed a similar technique to hurt people" (Lewis 2018).

In other words, the main motive to publish the study was a moral one, rather than a theoretical or technological one. Given that many countries still persecute homosexual people, Wang and Kosinski (2018, 248) considered it "critical to inform policymakers, technology companies and, most importantly, the gay community, of how accurate face-based predictions might be." Because of this information, individuals and companies could start taking preventive measures, researchers could design new protective technologies, and companies and governments could step up and create new policies and laws.

Most people would likely agree that it is best not to engage in research that could cause any kind of great harm. Issues only arise because people disagree on whether research will cause more harm than it will prevent. One way to deal with such dissensus is to construct a calculus and find out whether the moral benefits outweigh the costs. Wang and Kosinski clearly believe that the potential benefits of alerting the public and policy makers to the dangers of gaydar technology are greater than the potential harms associated with their paper. In principle, such warnings can be sensible. In an earlier paper, Kosinski indicated how Facebook "likes" can be used to predict many private traits, and he succeeded in urging Facebook to make these "likes" more invisible (Kosinski, Stillwell, and Graepel 2013). As to the risks, Wang and Kosinski believe their publication will not make much of a difference, as it is very easy to develop deep neural networks. All their tools and methods are widely known, and the data they used were all public.

However, we disagree with their self-assessment. In our view, Wang and Kosinski oversell the potential benefits while underestimating the risks.

First, the benefits depend on the likelihood that their paper significantly increased risk awareness. It is not clear, however, when such an increased risk awareness is significant and how this can be measured. Second, even though Wang and Kosinski may be right that their technology was already available, publishing a paper on this technology might have made that technology and its potential more conspicuous for ill-intentioned people and governments. Third, the (potential) surge in risk awareness has to increase the probability that the risk will be successfully mitigated. Ironically, Wang and Kosinski are quite pessimistic about this last point. They repeatedly emphasize that the rapid progress in artificial intelligence will inevitably erode our privacy, and there is not much we can do to protect ourselves:

> The digital environment is very difficult to police. Data can be easily moved across borders, stolen, or recorded without users' consent. Furthermore, even if users were given full control over their data, it is hard to imagine that they would not share anything publicly. Most people want some of their social media posts, blogs, or profiles to be public. Few would be willing to cover their faces while in public. . . . Consequently, we believe that further erosion of privacy is inevitable, and the safety of gay and other minorities who may be ostracized in some cultures hinges on the tolerance of societies and governments. (Wang and Kosinski 2018, 255–56)

Surely, LGBTQ activists are already aware that it is important to make societies and governments more tolerant. Therefore, it is not the case that Wang and Kosinski's paper made them realize that homonegativity is something that should be fought. More importantly, Wang and Kosinski's calculus may be wanting, but their critics' half-hearted weighing of costs and benefits is not much more compelling. The critics seem to assume that the benefits of the warning are nonexistent and that the risks are significant. Frankly, we are skeptical that a calculation of risks and benefits is even possible in this particular context. Some possible harmful and beneficial outcomes are known, but their probability is not (Hayenhjelm and Wolff 2012). In addition, we can hardly compare some of the moral costs and benefits. How exactly is one supposed to find a balance between one value (the protection of an underprivileged minority) and a quite unrelated value (scientific freedom and the duty to pursue truth)? After all, limiting the freedom of research in this case does not just limit the chances of creating awareness about the dangers of this technology (and acting on that awareness); it may also prompt other constraints on scientific inquiry. Again, we do not know the probability that such slippage will happen once constraints are placed on similar research. The bigger problem, however, is that even though we value both the rights of homosexual people and free scientific inquiry, we cannot easily determine how they trade off against each other.

Fortunately, the impossibility of a clear-cut consequentialist calculus does not mean that we cannot reflect fruitfully on the morality of Wang and Kosinski's research. A calculus is not necessary to understand that if the risks can be diminished without jeopardizing the benefits, it is irresponsible not to do so. A relevant implementation of this generally accepted rule can be found in cybersecurity research. There, vulnerabilities in security systems are studied and documented, but researchers initially do not publish their findings. They first inform the developer, vendor, or consumer about their findings, so that these stakeholders are given the opportunity to look for countermeasures. Once these stakeholders have sufficiently addressed the safety issues, the researchers share the vulnerabilities with a broader community of researchers and with stakeholders that are less directly involved (Arora, Telang, and Xu 2008).[7] Consequently, it may have been sensible for Wang and Kosinski to engage in their research project, since their results can help alert a vulnerable group to a particular danger and incentivize the development of countermeasures. It was less sensible, though, to publish their results in a high-profile journal and seek global media attention. To be fair, the authors took some steps to reduce the risk of misuse. For example, they did not publish the full source code and the learned data structures. Also, in an early research stage, they collaborated with stakeholders such as the American Civil Liberties Union. Still, we consider it rather irresponsible to share the results widely before preventive or mitigating measures were in place. By widely circulating their results, Wang and Kosinski increased the risks associated with their research without ensuring the development of preventive measures.[8]

Our cautious conclusion is that Wang and Kosinski's research was permissible, but it was probably impermissible to publish it the way they did. This conclusion still leaves a few important questions unanswered. We explore two of these questions in the following final section of this epilogue. The first question is whether the impermissibility of Wang and Kosinski's paper entails a more general impermissibility of publications on the etiology of homosexuality. The second question concerns the policing of the boundary between permissible and impermissible publications.

AGAINST CENSORSHIP

Kitcher argued for abstaining from research into cognitive differences among ethnic groups. Yet, he straightforwardly rejected any form of censorship of such research, because such censorship will often be counterproductive: "In a world where (for example) research into race differences in IQ is banned, the residues of the belief in the inferiority of the members of certain races are reinforced by the idea that official ideology has stepped in to conceal an uncomfortable truth" (Kitcher 2001, 105).

Kitcher is probably right in saying that censoring basic science can be counterproductive. Censoring applied research, however, is a different matter. The distinction is relevant here, because Wang and Kosinski themselves emphasize that their paper contributes very little to our theoretical knowledge. Primarily, what their publication offers is an innovative technological application. With regard to such applications, scholars rarely protest publication restrictions. When a newly developed nuclear technology can be used for military aims, most people think it is reasonable not to share details on how to build this innovative technology through easily accessible journal articles. Similarly, after the US anthrax attacks in 2001, the life sciences decided to restrict the circulation of research results that could be used to develop bioweapons. In 2004, the American National Research Council published *Biotechnology Research in an Age of Terrorism*, which tried to find a balance between freedom of scientific inquiry and national security. It recommended "relying on self-governance by scientists and scientific journals to review publications for their potential national security risks" (National Research Council 2004, 97; see also Kourany 2016).

A similar, widely supported appeal to self-regulation would be welcome in the social sciences, given the dangers posed by some applications of social science research. Self-regulation here does not mean that we should entirely leave it to the researchers themselves to decide whether publishing their results is warranted. After all, most researchers lack expertise to assess the dangers of their research and mitigate any risks. Moreover, controversial and groundbreaking topics often carry with them more scientific prestige, thus incentivizing scientists to do research like Wang and Kosinski's and share it as widely as they can. A few years after its publication, Wang and Kosinski's paper has already been cited hundreds of times, citations that provide the scientific credit researchers need to get job offers, tenure, and salary raises. Instead, the self-regulation we have in mind is similar to the self-regulation currently in place in cybersecurity and in the life sciences. Similar to these models, the regulation is done by scientists in scientific integrity committees and the like. Of course, most social science research is assessed by research ethics boards, and so was Wang and Kosinksi's research. Such ethics committees, however, currently focus primarily on protecting subjects that participate in a research study, and thus pay almost no attention to how the results of the study can be used by ill-intentioned people and governments. Research ethics boards rarely engage in a thorough analysis of the risks and benefits of the eventual circulation of the proposed research,[9] nor do they systematically give specific recommendations on how to prevent or mitigate the risks that the study can give rise to. If they did, many of the most pressing issues would be solved.

Should research ethics boards also put restrictions on research into the causes of sexual orientation? After all, such research can be instrumental for artificial intelligence that identifies sexual orientation or for prenatal screening techniques. On such techniques, Udo Schüklenk and colleagues write that "if prospective parents believe they are able to predict the sexual orientation of a fetus by using a prenatal screening technique, it is possible that they would choose to abort a fetus that seemed to be 'homosexually predisposed'" (Schüklenk et al. 1997, 9; see also Byne and Stein 1997). Although we think it is ultimately up to research ethics committees to make an informed decision on the desirability of specific research lines, there are reasonable grounds to be much more lenient toward basic research on the etiology of sexual orientation than toward applied research such as Wang and Kosinski's. First, there is Kitcher's argument that an imposed restriction on basic research is often counterproductive. Second, if we refrained from all new research on this topic, we would make it much more difficult to criticize the extant body of (fallible) research findings on this topic.[10] Third, if we decide to stop all research into the causes of sexual orientation because it underlies—or it might underlie—the development of potentially harmful technologies, we will have to limit an enormous series of research lines. In fact, it is not hard to fathom that almost all basic research in the social and life sciences can lead to findings that may undergird, either directly or indirectly, some subsequent morally undesirable technology.

Of course, it would be foolish to deny the harms research on sexual orientation have already caused, and only a little less foolish to deny that some research on homosexuality continues to be shaped by homonegative and heteronormative biases. Up to this day, these biases have both epistemic and moral costs. However, rather than forbidding all research we suspect to have morally questionable biases, we should document these past and present biases and study how they and their influence can be altered. And that is what we hope to have contributed to with this book.

Acknowledgments

For what it's worth, this book is the result of some fifteen years of collaborative thinking about science and homosexuality. In all these years, so many people have helped shape our ideas and arguments, either knowingly or unknowingly, that it is a rather hopeless task to list them all. Here we only mention those who would (rightly) come after us if we didn't.

Some have read and commented on earlier drafts of these chapters. In doing so, they kindly kept us from errors of all sorts: Luk Adriaens, Stijn Conix, John Corvino, Raf De Bont, Daniel Kelly, Olivier Lemeire, Paul Moyaert, Lesley Newson, Jacob Quick, Grant Ramsey, Jonathan Sholl, Karel Vereertbrugghen, and, of course, a colorful group of (more or less) anonymous reviewers we met along the way.

In matters more practical, we enjoyed the advice and help of Tristan Bates, Carl Demeyere, Stefan Derouck, Elizabeth Ellingboe, Keunho Hong, Eric Nenkia Bien, Diederik Walravens, and Sean Winkler.

In late 2015, our editor Karen Darling offered us an advance book contract, which we brazenly left lying around for four years. We thank her for her patience and perseverance and solemnly swear we will do better next time.

Parts of chapter 1 are based on a paper published in *Philosophy Compass* (De Block and Lemeire 2015). Parts of chapter 2 draw on a paper published in *Philosophy, Theory, and Practice in Biology* (Adriaens 2019). Parts of chapter 3 are based on papers published in *Philosophical Psychology* (De Block and Adriaens 2004) and *Perspectives in Biology and Medicine* (Adriaens and De Block 2006) and on a book chapter cowritten with Lesley Newson (Adriaens, De Block, and Newson 2012). Parts of chapter 4 draw on a review paper for the *Journal of Sex Research* (De Block and Adriaens 2013) and on a chapter on conceptual engineering in philosophy of medicine in the edited volume *Explaining Health across the Sciences* (De Block and Sholl 2020). The last sections of the epilogue build on a paper cowritten by Stijn Conix (De Block and Conix, forthcoming). Parts of many chapters draw on a Dutch book on science and homosexuality published by LannooCampus (Adriaens and De Block 2015).

Notes

INTRODUCTION

1. The question at hand is question 26 in book 4 of the *Problemata*, quoted from the Loeb Classical Library's translation (Aristotle 2011a). We have translated all non-English quotations ourselves, unless there is a good translation available, in which case we mention the translator(s).

2. We borrow the term "gay science" from Timothy Murphy's eponymous 1997 book on the morality of sexual orientation research (Murphy 1997). The terms "gay studies" and "queer studies" are less fit for our purposes, as they focus, or at least did originally, on a more limited set of sciences, especially history and literary theory.

3. Scientific disputes are often thought to end when one party is proved right, but the history of science shows some alternative arrangements. Sometimes both sides are proved wrong; sometimes one side is silenced by the church or some worldly authority; sometimes one of the influential scholars dies or loses funding. All these twists of history could have been different, and the same is true for their outcomes.

4. To be fair, the quote is often attributed to Feynman, but "whilst Feynman did nay-say philosophy quite a bit, it's not so clear that he did actually say those words" (Trubody 2016, 10).

5. Fostering dialogue among the sciences has the added advantage of revealing or emphasizing the multiplicity or multidimensionality of homosexuality. Like so many other natural phenomena, homosexuality has several dimensions that more or less correspond to the many sciences involved in studying it. The life sciences, e.g., anatomy and genetics, focus mainly on the biological causes of homosexuality. Human sciences, e.g., social psychology and anthropology, study interindividual and intercultural differences in homosexual behaviors, desires, preferences, orientations, and identities. Dissensus often arises when scientists disregard this multidimensionality, for example, by overestimating the importance of the dimension they study (Daston 2010).

6. Some scientists make a further distinction between modern (male) homosexuality and what they describe as "transgendered androphilia"—a homosexual phenotype identifiable by its marked effeminacy (VanderLaan, Ren, and Vasey 2013). In our view, however, the distinction between "transgendered" and "sex gender congruent" homosexuality runs through the diverse forms of homosexuality that we described earlier. For more on this, see chapter 3.

7. Among the Western populations that have been studied, bisexual patterns of attraction are rarer in men than in women, and exclusively homosexual attractions

are rarer in women than in men. The reasons for this sex difference are not known. It might reflect prenatal biological sex differences; it might reflect the influence of cultural factors that create different social contexts for the development of female versus male same-sex sexuality; it might reflect differences in women's and men's susceptibility to such social influences (Baumeister 2000); or it might simply result from measurement error (Bailey et al. 2016, 54).

8. For various reasons, we prefer the term "homonegativity" to the old "homophobia." First, the term "homophobia" suggests it is a mental disorder, much like other phobias. However, most researchers consider homonegativity to be a political and sociological problem, rather than a medical condition. Second, the suffix "phobia" seems to indicate that the negative attitudes are primarily characterized or fueled by fear. Nobody denies that fear can play a role in homonegativity, but research indicates that many other emotions, such as disgust and anger, are also involved, and may even be more fundamental to homonegativity than fear (Herek 2004; Olatunji 2008).

9. It is unclear whether Tennent meant his complaint to be tongue-in-cheek. Zuk (2011, 130) believes he did; Bagemihl (1999, 89) clearly does not. Tennent's own contribution to the literature consisted in a description of courtship behavior between male butterflies in the presence of willing females. He describes his horror and indignation and his concern about the survival of the butterfly colony and makes the inevitable comparison with human homosexuality (Tennent 1987).

CHAPTER 1

1. All quotes are from the Loeb Classical Library's translation of the *Problemata* (Aristotle 2011a, 2011b). Hereafter, citations of the *Problemata* are given parenthetically in-text with reference to particular books.

2. In the 1940s, castration was often ordered without consent (Plant 2011). Interestingly, the Nazis had comparatively little interest in female homosexuals.

3. Recent estimates of the concordance for schizophrenia among identical twins are much lower (Hilker et al. 2018).

4. See, e.g., the websites for Joseph Nicolosi, https://www.josephnicolosi.com/, and Core Issues Trust, https://www.core-issues.org/.

5. Importantly, the 2019 genome-wide association study operationalized same-gender sexual behavior as "ever" engaging in same-sex sexual behavior (Ganna et al. 2019). As Lisa Diamond (2021, 2) notes, "Such categorization conflicts with scientific and lay conceptions of sexual orientation as an enduring and overarching sexual predisposition for one or more genders."

6. Scott Alexander (2016) identifies many misconceptions about the value of unshared environment measures.

7. Interestingly, the statistical techniques used for heritability studies entail that interactions between parenting style and genes would show up as a genetic effect, not as an effect of the nonshared environment (Polderman et al. 2015).

8. There is some debate within philosophy on whether heritability tells us something about an individual. Most philosophers think that it tells us absolutely nothing about the individual. Neven Sesardic (2005, 55) disagrees: "If the broad heritability of a trait is high, this does tell us that any individual's phenotypic divergence from the mean is probably more caused by a non-standard genetic influence than by a non-typical environment."

9. Often, this sort of developmental robustness is taken to be sufficient for canalization (McCarthy et al. 2015).

10. It may even entail that sexuality can be changed by interventions that precede this critical period. A recent editorial on Blanchard's research on antigens and the birth order effect asks, "If an antibody response can be confirmed as an etiological factor, might primary prevention interventions be possible?" (Rao and Andrade 2019, 110).

11. They do not *necessarily* commit this or another fallacy, though, since it is in this case logically valid to use the consequent as nondecisive evidence for the antecedent. If innateness implies immutability, the immutability of a trait provides some evidence for the trait's innateness.

12. Others have made different distinctions. Sally Haslanger's (2003) is perhaps the most influential classification of constructivism variants, which distinguishes between causal and constitutive constructivism. We agree with Teresa Marques (2017) that this distinction is not always useful, however, because every constitutively constructed X is always also causally constructed. That said, we do think that causal constructivism comes very close to—or might even be identical to—what we call ontological constructivism.

13. This is George Chauncey's (1982, 116) formulation of epistemological constructivism: "'Sexual inversion' referred to a broad range of deviant gender behavior, of which homosexual desire was only a logical but indistinct aspect, while 'homosexuality' focused on the narrower issue of sexual object choice. The differentiation of homosexual desire from 'deviant' gender behavior at the turn of the century, reflects a major reconceptualization of the nature of human sexuality, its relation to gender, and its role in one's social definition." This quote suggests that Chauncey is not a semantic social constructivist: he contends that homosexual desire was captured by "sexual inversion."

14. Some social constructivists, such as David Halperin and Jeffrey Weeks, do make the necessary distinctions. Some critics do also distinguish different types of constructivism. John Thorp (1992, 56), for example, distinguishes between a weak (epistemological) and a strong (ontological) version of social constructivism: "The weak form is that different people naturally have a whole array of different sexual tastes and desires; what we have done is to categorize and label these in such a way that the great divide is established upon the gender of the object of desire, rather than upon its shape, size, vigour, colour, or social class. We have drawn the conceptual lines, and now are puzzled by them. The stronger form of the thesis is that the desires themselves have been socially produced."

15. Admittedly, even this finding can be squared with ontological essentialism. For example, one can still maintain that there is only one kind of homosexuality and that this kind is defined not by its biological causes but by its phenotype.

16. Daniel Ortiz (1993, 1835) makes a similar claim: "Much of the debate's intractability, in fact, stems from confusing fundamentally different types of claims. In a sense, even the participants misunderstand what the debate is about." However, we think he also misunderstands (part of) the debate, because he goes on to state that the debate is not "even partly about the causes of homosexuality."

CHAPTER 2

1. The first volume of *Lampyrid: The Journal of Bioluminescent Beetle Research*, published in 2011, contained an English translation of Peragallo's paper. Unfortunately, this translation is too liberal for our purposes.

2. One can indeed distinguish between male and female cockchafers by their antennae, as male antennae are slightly bigger and longer.

3. Brooks (2009, 153) translates the French word "goût" as "choice," which is definitely a mistake. "Goût" can have multiple meanings, including "taste" and "manner," both of which do not seem to be relevant in this context. However, "goût" can also mean "appetite" in a variety of contexts, including gastronomy and sex. "Avoir du goût pour quelqu'un," for example, can be translated as "to feel (sexually) attracted to someone." For Gadeau de Kerville, the word "goût" obviously refers to sexual lust and even sexual preference, as he himself admitted when juxtaposing "par goût" and "par préférence" (Gadeau de Kerville 1896a, 7).

4. In the 1950s, the term "pederasty" was still fashionable in the scientific literature, as in, for example, Clellan Ford and Frank Beach's *Patterns of Sexual Behavior* (1951, 126). Here it also referred to homosexuals who engage in anal sex.

5. Féré would criticize Gadeau de Kerville on this issue. While conceding that masturbation-like behaviors occur in some animal species (he even mentions a colleague who had "accused the camel and the elephant" of masturbating [Féré 1899, 75]), he emphasized that they always arose under special circumstances, to wit "a local irritation" (76).

6. According to Féré, congenital homosexuality is an incurable disease, and all efforts to "convert" homosexuals to becoming heterosexual do more damage than good: "Congenital sexual inversion is beyond the range of medicine. It is no more possible to restore the sexual sense of a congenital invert than to restore color vision in a daltonist.... Since inverts are degenerates, and since they are able to produce a pathological offspring if they are trained (or rather converted) with success, they should be advised to live outside of marriage" (Féré 1896, 9).

7. The *Melolontha* story partly confirms and partly contradicts Bagemihl's analysis of the history of research on animal homosexuality. The debate between Gadeau de Kerville and Féré teaches us, for example, that there has always been a variety of explanatory hypotheses, even in the very first scientific debates on animal homosexuality. Moreover, gay activists such as Karl Heinrich Ulrichs quickly picked up on some of the early findings and eagerly quoted them as evidence of the naturalness and normality of human homosexuality. So it would perhaps be more fair to say that animal homosexuality also played a positive part in the history of biology.

8. Of course, there were exceptions. Clement of Alexandria thought that the hyena was "innately homosexual," as he (erroneously) believed the animal to have an orifice specifically designed for homosexual activities (quoted in Boswell 1980, 156n81).

9. The full passage in Plutarch's *Gryllus* reads: "A cock that mounts another for lack of a female is burned alive because some prophet or seer declares that such an event is an important and terrible omen" (Plutarch 1900, book IX, 990E; see also Karsch 1900, 132).

10. However, the concept of subjective sexual arousal is notoriously poorly defined in the literature.

11. Stein's dismissal of avian intelligence is too quick. After all, there are far more intelligent bird species than there are intelligent mammal species (Isler and Van Schaik 2009).

12. In philosophical parlance, one can distinguish desires that are homosexual *de re* and desires that are homosexual *de dicto*. In the first case, sexual desires are directed to individuals who happen to be of the same biological sex; in the second case, sexual desires are directed to individuals *because* they are of the same biological sex (Markie

and Patrick 1990). Sexual desires that are homosexual desires *de re* can be further subdivided into two categories based on whether the desiring individual "knows" that the desired individual is of the same biological sex. However, even those desires that are homosexual *de dicto* are not necessarily identical to a homosexual preference.

13. This view of sexual orientation remains quite close to the kind of desire dispositionalism that was developed by Esa Díaz-León, which holds that for a man to be sexually oriented toward men he must be disposed to sexually desire men in a rather wide variety of circumstances (see Díaz-León, forthcoming). Other so-called social metaphysicians have argued that the relevant disposition is not a disposition to desire but a disposition to behave (see, e.g., Dembroff 2016).

14. The British comedian Ricky Gervais discusses some of Bagemihl's illustrations in his comedy show *Animals*. A drawing of two male stump-tailed macaques (*Macaca arctoides*) in mutual fellatio (Bagemihl 1999, 313) brings him to criticize Bagemihl's decision to illustrate his book with line drawings rather than photographs: "I will say one thing about these [drawings]: they are not photographs! That's not proof, is it? Someone just drew them. He [Bagemihl] must have gone to a publisher and went: [Bagemihl:] 'I've written a brilliant book. All animals are gay.' [Publisher:] 'They are what?' [Bagemihl:] 'Animals are as bent as a nine bob note.' [Publisher:] 'Wow, that's amazing. Have you got pictures? Can I see them? Tomorrow?' . . . It's not proof, is it? A photograph's a proof. . . . Those little stump-tailed macaques are probably getting beaten up in the jungle, while they're going: 'That's not us! Anyone can draw that!'"

15. There is some debate on how accurate the mirror test is for establishing the presence of self-awareness. It seems that this test results in quite a few false positives and false negatives (Cammaerts 2017; Cazzolla Gatti 2016).

16. The full quote reads as follows: "[Odenwald] and Zhang, with no apparent sense of irony or humor, claimed that the mutant fruit flies were 'gay' and, moreover, that genetic manipulations made them that way. . . . [Odenwald] and Zhang did not hesitate to classify the mutant flies in terms of identity—that is, as gay fruit flies rather than simply as flies that exhibited homosexual behavior" (Terry 2000, 167).

17. There are many other reasons why Kelch's explanation falls short. First, it was challenged by later observations of similar couplings where some penetrated males were actually larger than the penetrating males (Doebner 1850; Osten-Sacken 1879). Furthermore, Kelch simply assumed that the penetrated partner had been unwilling and therefore needed to be conquered—a situation in which the penetrator's power proved useful. This assumption fits into what Bagemihl considers to be a broader strategy in the history of research on animal homosexuality: "Same-sex activity is routinely described as being 'forced' on other animals when there is no evidence that it is, and a whole range of 'distressful' emotions are projected onto the individual who experiences such 'unwarranted advances'" (Bagemihl 1999, 90). As explained earlier, Kelch's critics also debunked his interpretation as they reasoned that copulation and penetration in *Melolontha* males would be technically impossible without the cooperation of both partners (Osten-Sacken 1879).

18. Gray's focus on pleasure is also problematic because some sexual activities are accompanied by *un*pleasurable arousal. The fact that some people experience anal penetration as painful, for example, does not mean that it cannot be a sexual activity. Conversely, some human activities are aimed at pleasurable arousal, even though they are not considered sexual. Most people would say, for example, that the enjoyment of a lavish dinner does not involve sexual pleasure (Soble 2008).

19. While some scientists explicitly claim that diddling can be found only in baboons and not in humans, others have pointed to similar behaviors documented in anthropological studies of the Bedamini in Papua New Guinea and Indigenous Australians, which also appear to be functionally similar to diddling (Smuts and Watanabe 1990, 169; Bagemihl 1999, 681–82n70). Part of the problem is due to the fact that biologists have found it very difficult, if not impossible, to conceptualize and operationalize "similarity" in univocal ways (Ghiselin 2005). As a result, it is not clear as to when we would consider two tokens of a type of (sexual) behavior to be similar. One could argue, for example, that human diddling should not be considered similar to baboon diddling because the behaviors serve different functions in their respective species. It is possible, however, to think of an equally plausible argument in favor of the opposite position, i.e., that human and baboon diddling are similar because they both (supposedly) originate in social learning or because humans and baboons share a common ancestor that also practiced diddling in a homosexual context.

20. Presumably, there are other examples. Bagemihl (1999, 15) mentions the case of male ostriches: "Perhaps most interesting are those creatures that have a special courtship pattern found only in homosexual interactions. Male ostriches, for example, perform a unique 'pirouette dance' only when courting other males."

21. Bagemihl aptly remarks, however, that some biologists seem to use two different definitions of sexuality: a broad one in discussions of heterosexual behavior and a very narrow one in discussions of homosexual behavior: "For example, simple genital nuzzling of a female Vicuna by a male—taking place outside of the breeding season, and without any mounting or copulation to accompany it—is classified as sexual behavior, while actual same-sex mounting in the same species is considered nonsexual or 'play' behavior" (Bagemihl 1999, 117).

22. A final example of this objection to humanizing animal sexuality can be found in Stein's critique of (what he thinks to be) a representative example of research on homosexuality in fruit flies: "According to Burr . . . , the behavior exhibited by fruit flies with the fruitless gene is 'a dramatic example of homosexuality in animals.' What Burr is actually discussing amounts to a dramatic example of how some of the biological literature on animal sexual behavior . . . is guilty of extreme anthropomorphism" (Stein 1999, 166).

23. The context of the interview suggests that Bagemihl was primarily interested in the legal and moral implications of his research topic rather than in its repercussions on scientific research on human homosexuality. In 2003, his work was cited in the US Supreme Court case *Lawrence v. Texas* to demonstrate the naturalness of male homosexuality. The law in question, which criminalized sodomy, was ruled unconstitutional.

24. To be clear, Fausto-Sterling is not arguing that the penetrating partner's behavior is necessarily a better model for human homosexuality. (In fact, one could argue that the error hypothesis is a better explanation for such behavior.) She is criticizing what she sees as a common and mindless assumption: the idea that male homosexuality is always linked to passivity and femininity.

25. To be fair, Beach anticipated this critique by noting that some animal scientists tend to mix up two rather different meanings of the term "homosexual," even though the Aristotelian author mentioned in our general introduction already neatly distinguished between both. On the one hand, "homosexual" refers to individuals whose mating behavior resembles that of the opposite sex; on the other hand, it refers

to individuals who exhibit sexual behavior that is typical of their sex but as a reaction to an individual of the same sex (Beach 1948).

26. Prenatal hormonal exposure is mostly measured indirectly by investigating the variation in certain characteristics that are known to correlate well with prenatal exposure to certain hormones.

CHAPTER 3

1. In a paper presented at the famous Cambridge Conversazione Society on December 8, 1894, Moore made the following confession: "When I came up to Cambridge, I did not know that there would be a single man in Cambridge who fornicated; and, till a year ago, I had no idea that sodomy was ever practiced in modern times. My discoveries on these points have naturally brought the subject very much before my mind, and perhaps made me attach an undue importance to it" (quoted in Regan 1986, 39). It is more likely, however, that Hutchinson was already familiar with homosexuality prior to his Cambridge years, as homosexuality was also very common at English public schools, including Hutchinson's own Gresham School (Slack 2010).

2. In one of his Marginalia columns, Hutchinson uncharacteristically lashed out at the statistical stance of much midcentury research on homosexuality, like Ford and Beach's *Patterns of Sexual Behavior* (1951). According to Hutchinson (1953, 69): "It is extremely curious to see that a generation of functionalist anthropologists ... should give place to a statistical generation whose members are now taking tiny bits of sexuality out of context without the slightest shame simply because they have an IBM machine rather than a classical education to help them." He much preferred Margaret Mead's "functionalist" work on the topic (103).

3. Most intelligence researchers now think that the correlation between intelligence and fertility is indeed negative, but that this effect is mostly mediated by education. More education leads to both lower fertility and higher intelligence (Meisenberg 2010).

4. Evolutionary biologist Michel Raymond and his colleagues found highly similar results in a population of male homosexuals in Indonesia (Nila et al. 2018).

5. Some historians argue that some forms of early modern homosexuality were already more egalitarian than others. It is known, for example, that same-sex marriages occurred in the 1500s in Renaissance Rome and early modern France (Bray 2003; Tulchin 2007).

6. Biologists are more cautious when it comes to female homosexuality, acknowledging that it may not affect reproductive success: "It is not clear whether homosexuality in females has entailed a nontrivial fecundity reduction in the evolutionary past, as opposed to present times" (Camperio-Ciani, Battaglia, and Zanzotto 2015, 3).

7. Some evolutionists seem to forget that this logic does not work for homosexual behaviors (for reasons we discuss in the next section). In a recent paper, for example, Colledani and Camperio-Ciani (2021, 4) erroneously claim that "homosexuality ... involves sexual behaviors that are nonreproductive and that, *consequently*, should reduce the reproductive fitness" (italics ours).

8. We disagree with Sommer's allegation that the alliance formation hypothesis would be "so ubiquitous as to be untestable" (Sommer, Schauer, and Kyriazis 2006, 267) because its predictions are unclear. Curiously, many of the chapters in *Homosexual Behaviour in Animals* (Vasey and Sommer 2006) do support the alliance formation hypothesis.

9. Tovee et al. (2006) claim that learning steers the norms of attractiveness. If that is the case, and if the learning is of a social kind, then the preferences are culturally shaped.

However, it is also possible that norms of attractiveness respond to environmental variation without any direct involvement of social learning. When it's warm, we tend to be more thirsty than when it's cold, and that environmentally triggered variation in our preference is not cultural.

CHAPTER 4

1. As we explained in chapter 1, many constructivists believe that Krafft-Ebing not only captured this feature but also shaped or even created it.

2. *Psychopathia sexualis* is somewhat inconsistent on this point, since a desire to rape is also sometimes described as a core symptom of sadism and hyperesthesia (see also Knight and Prentky 1990).

3. For Freud, the baseline normal sexuality is heterosexuality because of its connection to reproduction. Still, in *Three Essays* (Freud 1960b), he attempted to debunk the notion that humans are programmed for heterosexuality. Confusingly, later psychoanalysts defended that very notion, as we mention later in this chapter, and they found some ammunition for this view in Freud's later works. There is little doubt that, at some point in his career, Freud did indeed consider heterosexuality as some kind of an ideal, but given the role of repressed sexuality in the etiology of psychopathology more generally, he would probably not have thought that an exclusively heterosexual orientation was always best. Admittedly, however, he never goes as far as to say that exclusive heterosexuality is a perversion.

4. The physician and psychologist Henry Havelock Ellis supported Hirschfeld on this point. He also argued that sexology played a pioneering role in the political emancipation of homosexuals and other sexual minority groups (Weeks 2000). Ellis often took the lead in this activism, partly due to his lavender marriage to the openly lesbian writer Edith Lees.

5. The most important study in this tradition was Clelia Duel Mosher's survey conducted and published in 1892 (Kahan 2021). Much of this statistical research was based on questionnaires. Some of these studies were used as the scientific core of best-selling popular science books, like Katherine B. Davis's *Factors in the Sex Life of Twenty-Two Hundred Women* (1929). R. L. Dickinson and L. Beam's *A Thousand Marriages* (1931) was quite similar and based on the data that Dickinson gathered as a practicing gynecologist.

6. Marmor could also have mentioned Freud's speculations about bisexuality (Sulloway 1979), but he referenced only the work of the American ethologist Frank Beach, who coauthored the classic *Patterns of Sexual Behavior* in 1951. Here the authors indeed referred to "the bisexuality of the physiological mechanisms for mammalian mating behavior" and a "fundamental mammalian heritage of general sexual responsiveness" (Ford and Beach 1951, 258–59).

7. Krafft-Ebing (1886, 501) once noted that "the nature of the act can never, in itself, determine a decision as to whether it lies within the limits of mental pathology.... The perverse act does not per se indicate perversion of instinct."

8. It is an understatement to call Levin a controversial philosopher. He is a suspected White supremacist and wrote pieces on race for *American Renaissance*, a White supremacist magazine.

9. Interestingly, Spitzer repeatedly recommended later editors of the *DSM* to follow Wakefield's account of disorder, particularly his evolutionary account of dysfunction

(Spitzer and Wakefield 1999). His enthusiasm, however, did not catch on, at least judging from Michael First's later struggle with the concept of dysfunction.

10. When polled on this issue, just a little over 50 percent of the philosophers answered that they accept or lean toward moral realism (Bourget and Chalmers 2014).

11. To be clear, conversion therapy is not a subtle way of controlling sexuality.

12. Cooper's statement that a disorder is a bad thing to have is part of her analysis of the concept of disorder. Here we refer to her normativist position simply to underscore the connection between disorder and undesirability. This connection is the plausible starting point of Cooper's analysis. However, we acknowledge that there is a long argumentative way to go from this starting point to Cooper's conclusion that this undesirability defines disorder.

13. In a critique of normativist analyses of disorder, Boorse provides a good example of this strategy. In his view, these analyses entail that an unwanted pregnancy is a disorder, as it is undesirable and unlucky, which to him seems an absurd implication: "To call pregnancy per se unhealthy would strike at the very heart of medical thought; it is the analytic equivalent of the 'Game Over' sign in a video game" (Boorse 1997, 44).

14. Quill R. Kukla (2019) offers an interesting example when proposing a pragmatic analysis of infertility as a disorder. They suggest carefully weighing the various values and ideological interests involved as well as the assumptions about these values and interests.

15. Maybe our beliefs and attitudes have evolved so substantially over the past five decades that we now see homosexuality as a prototypically healthy condition. Still, in 1974 homosexuality really was a controversial condition, and our claim is that even back in 1974, it should have been treated as a prototypically healthy condition, not on scientific but on moral grounds.

16. There are not many polls that track these changes. However, both in Germany and in France, the public has been polled a couple of times on how they conceived of homosexuality. It is interesting to see that in Germany the percentage of people who considered homosexuality a vice halved between 1949 and 1976. In 1949, 53 percent of the respondents saw it as a vice, whereas that number dropped to less than 25 percent in 1976. In that same period, more and more people started to consider homosexuality as a disorder. That number went up from 39 percent to almost 50 percent (De Boer 1978). This can be considered (weak) evidence for the claim that moral (and legal) judgment was replaced by medical judgment.

EPILOGUE

1. As John Corvino (2013) explains, even those who reject Hume's critique and adhere to natural law theory do not always condemn homosexuality. For instance, they may argue that homosexuality simply does not have a procreative function, and hence we "should not fault same-sex sex for failing to procreate" (Blankschaen 2019, 433).

2. We could also deduce all kinds of other information from a preference for pink. After all, if the preference for pink is caused by an alleged homosexual essence, then we could deduce from that concrete preference other properties that the homosexual will have, based on his essence.

3. Here, Haslam and Levy rely on attribution theory, a theory that asserts that when a person's disvalued condition is attributed to forces outside one's control, the person is less likely to be held responsible for that condition. It should be kept in mind that this application of attribution theory means that homosexuality is judged less negatively

and not that essentialism leads to a positive attitude. After all, attribution theory also predicts that we will be less positive about a quality that we value that appears to be biologically fixed.

4. Kitcher (2001, 98) puts it slightly differently: "The bias in favor of the hypothesis [that Blacks are naturally unsuited for a particular role] is so strong that most members of the society will take evidence that, when assessed by the most reliable methods, would yield a probability for the hypothesis of roughly 0.5 to confer a probability close to 1 on the hypothesis."

5. To counter this kind of essentialist homonegativity, a virtual (though now mostly inactive) gathering place was created, called QueerByChoice.com. There, the idea that gay people "cannot help" being homosexual is rejected and replaced by the view that gay people should value and embrace their sexual orientation as the result of many indirect choices. QueerByChoice users even speculate that "anyone who's already discovered the joys of same-sex attraction could hardly be expected to ever develop much interest in dealing with all the inequalities and communication difficulties of opposite-sex attraction." "Could You Choose to Turn Hetero If You Wanted To?," QueerByChoice.com, http://www.queerbychoice.com/turnhetero.html.

6. According to A. Gelman, Mattson, and Simpson (2018), more than half the people identified by Wang and Kosinski's artificial gaydar as homosexual are in fact not homosexual at all.

7. Similarly, there is a lot of research on the genomics of deadly viruses, but the viruses' genomes are not made publicly available because there is a real danger that terrorists would use this information to develop effective bioweapons (Kourany 2016).

8. Given Wang and Kosinski's own apparent skepticism regarding the possibility of effective countermeasures, it may even be considered a questionable practice to engage in this type of research.

9. A shallow analysis does not always lead to a lenient attitude regarding the permissibility of the proposed research. In fact, many have complained that research ethics committees have encumbered the social justice possibilities of research on vulnerable populations (Detamore 2016).

10. Related to this, we doubt that new research into the causes of sexual orientation will really be game-changing for potentially harmful technologies. If anything, recent research advances in the etiology of homosexuality have shown that this etiology is much more complex than was thought at the end of the twentieth century, with this complexity related to both biological and nonbiological pathways as well as difficulties with the explanandum, i.e., homosexuality, and how to define it (Balthazart 2011a).

References

Abelove, Henry, Michèle Aina Barale, and David M. Halperin, eds. 1993. *The Lesbian and Gay Studies Reader*. London: Routledge.

Adriaens, Pieter R. 2019. "In Defence of Animal Homosexuality." *Philosophy, Theory, and Practice in Biology* 11 (22): 1–19.

Adriaens, Pieter R., and Andreas De Block. 2006. "The Evolution of a Social Construction: The Case of Male Homosexuality." *Perspectives in Biology and Medicine* 49 (4): 570–85.

———. 2015. *Born This Way: Een Filosofische Blik op Wetenschap en Homoseksualiteit*. Tielt, Belgium: LannooCampus.

———. 2016. "Lesser-Known Theories of Homosexuality." In *Encyclopedia of Evolutionary Psychological Science*, edited by Todd K. Shackelford and Viviana A. Weekes-Shackelford, 1–9. Cham, Switzerland: Springer.

Adriaens, Pieter R., Andreas De Block, and Lesley Newson. 2012. "Evolutionary Theory, Constructionism and Male Homosexuality." In *Sex, Reproduction and Darwinism*, edited by Filomena de Sousa and Gonzalo Munevar, 77–94. London: Pickering & Chatto.

Ainley, David G. 1978. "Activity Patterns and Social Behavior of Non-Breeding Adelie Penguins." *The Condor* 80 (2): 138.

Akankwatsa, Patricia. 2019. "Homosexuality Not Genetic, Says Study." *Independent*, September 10. https://www.independent.co.ug/homosexuality-not-genetic-says-study116416-2/.

Alexander, Scott. 2016. "Non-Shared Environment Doesn't Just Mean Schools and Peers." *Slate Star Codex* (blog), March 16. https://slatestarcodex.com/2016/03/16/non-shared-environment-doesnt-just-mean-schools-and-peers/.

Allison, Anthony C. 1954. "Notes on Sickle-Cell Polymorphism." *Annals of Human Genetics* 19 (1): 39–51.

Allport, Gordon. 1954. *The Nature of Prejudice*. Reading, MA: Addison-Wesley.

Altemeyer, Bob. 1996. *The Authoritarian Specter*. Cambridge, MA: Harvard University Press.

Anagnostou-Laoutides, E. 2015. "Luxuria and Homosexuality in Suetonius, Augustine, and Aquinas." *Mediaeval Journal* 5:1–32.

Anderson, Benedict. 1983. *Imagined Communities: Reflections on the Origin and Spread of Nationalism*. London: Verso.

Andrews, Kristin. 2014. *The Animal Mind: An Introduction to the Philosophy of Animal Cognition*. New York: Routledge.

Ankeny, Rachel, and Sabina Leonelli. 2020. *Model Organisms*. Cambridge: Cambridge University Press.

Anonymous. 1991. "Expert Says Mama Sheep Makes Ram a Stud or Dud." *The Spokesman Review*, June 16.

APA (American Psychiatric Association). 1952. *Diagnostic and Statistical Manual: Mental Disorders*. Washington, DC: American Psychiatric Association.

———. 1968. *Diagnostic and Statistical Manual of Mental Disorders*. 2nd ed. Washington, DC: American Psychiatric Association.

———. 1980. *Diagnostic and Statistical Manual of Mental Disorders*. 3rd ed. Washington, DC: American Psychiatric Association.

———. 1987. *Diagnostic and Statistical Manual of Mental Disorders*. 3rd rev. ed. Washington, DC: American Psychiatric Association.

———. 1994. *Diagnostic and Statistical Manual of Mental Disorders*. 4th ed. Washington, DC: American Psychiatric Association.

———. 2000. *Diagnostic and Statistical Manual of Mental Disorders*. 4th rev. ed. Washington, DC: American Psychiatric Association.

Apostolou, Menelaos. 2010. "Bridewealth as an Instrument of Male Parental Control over Mating: Evidence from the Standard Cross-Cultural Sample." *Journal of Evolutionary Psychology* 8:205–16.

Ariew, Andre. 1996. "Innateness and Canalization." *Philosophy of Science* 63:S19–27.

———. 1999. "Innateness Is Canalization: A Defense of a Developmental Account of Innateness." In *Biology Meets Psychology: Conjectures, Connections, Constraints*, edited by Valerie Hardcastle, 117–38. Cambridge, MA: MIT Press.

Aristotle. 2011a. *Problems*. Vol. 1, *Books 1–19*. Translated by Robert Mayhew. Loeb Classical Library 316. Cambridge, MA: Harvard University Press.

———. 2011b. *Problems*. Vol. 2, *Books 20–38*. Translated by Robert Mayhew. Loeb Classical Library 317. Cambridge, MA: Harvard University Press.

Arora, A., R. Telang, and H. Xu. 2008. "Optimal Policy for Software Vulnerability Disclosure." *Management Science* 54:642–56.

Bagemihl, Bruce. 1999. *Biological Exuberance: Animal Homosexuality and Natural Diversity*. New York: St. Martin's Press.

Bagley, Christopher, and Pierre Tremblay. 1998. "On the Prevalence of Homosexuality and Bisexuality in a Random Community Survey of 750 Men Aged 18 to 27." *Journal of Homosexuality* 36:1–18.

Bailey, J. Michael, Paul L. Vasey, Lisa M. Diamond, S. Marc Breedlove, Eric Vilain, and Marc Epprecht. 2016. "Sexual Orientation, Controversy, and Science." *Psychological Science in the Public Interest* 17 (2): 45–101.

Bailey, J. Michael, and Kenneth J. Zucker. 1995. "Childhood Sex-Typed Behavior and Sexual Orientation: A Conceptual Analysis and Quantitative Review." *Developmental Psychology* 31 (1): 43–55.

Bailey, Nathan, and Marlene Zuk. 2009. "Same-Sex Sexual Behavior and Evolution." *Trends in Ecology & Evolution* 24:439–46.

Bakker, Julie, Teus Brand, Jan van Ophemert, and A. Koos Slob. 1993. "Hormonal Regulation of Adult Partner Preference Behavior in Neonatally ATD-Treated Male Rats." *Behavioral Neuroscience* 107 (3): 480–87.

Balthazart, Jacques. 2011a. "Minireview: Hormones and Human Sexual Orientation." *Endocrinology* 152:2937–47.

———. 2011b. *The Biology of Homosexuality*. Oxford: Oxford University Press.

———. 2018. "Fraternal Birth Order Effect on Sexual Orientation Explained." *Proceedings of the National Academy of Sciences* 115 (2): 234–36.
Barnes, Elizabeth. 2016. *The Minority Body: A Theory of Disability*. Oxford: Oxford University Press.
Barnett, Michael D., Haluk C. M. Öz, and Arthur D. Marsden. 2018. "Economic and Social Political Ideology and Homophobia: The Mediating Role of Binding and Individualizing Moral Foundations." *Archives of Sexual Behavior* 47 (4): 1183–94.
Barron, Andrew B., and Brian Hare. 2019. "Prosociality and a Sociosexual Hypothesis for the Evolution of Same-Sex Attraction in Humans." *Frontiers in Psychology* 10:2955.
Barthes, Julien, Pierre-André Crochet, and Michel Raymond. 2015. "Male Homosexual Preference: Where, When, Why?" Edited by Garrett Prestage. *PLOS One* 10 (8): e0134817.
Bartoš, Ludek, and Jana Holečková. 2006. "Exciting Ungulates: Male-Male Mounting in Fallow, White-Tailed and Red Deer." In *Homosexual Behaviour in Animals: An Evolutionary Perspective*, edited by Volker Sommer and Paul L. Vasey, 154–71. Cambridge: Cambridge University Press.
Baumeister, R. F. 2000. "Gender Differences in Erotic Plasticity: The Female Sex Drive as Socially Flexible and Responsive." *Psychological Bulletin* 126 (3): 347–74.
Bayer, Ronald. 1981. *Homosexuality and American Psychiatry: The Politics of Diagnosis*. Princeton, NJ: Princeton University Press.
———. 1987. "Politics, Science, and the Problem of Psychiatric Nomenclature: A Case Study of the American Psychiatric Association Referendum on Homosexuality." In *Scientific Controversies: Case-Studies in the Resolution and Closure in Science and Technology*, edited by H. Tristam Engelhardt Jr. and Arthur Caplan, 381–400. Cambridge: Cambridge University Press.
Beach, Frank. 1948. *Hormones and Behavior*. New York: Hoeber.
Beccalossi, Chiara. 2010. "Nineteenth-Century European Psychiatry on Same-Sex Desires: Pathology, Abnormality, Normality and the Blurring of Boundaries." *Psychology & Sexuality* 1:226–38.
Bell, Alan, and Martin Weinberg. 1978. *Homosexualities: A Study of Diversity among Men and Women*. London: Mitchell Beazley.
Berenbaum, S. A., and A. M. Beltz. 2011. "Sexual Differentiation of Human Behavior: Effects of Prenatal and Pubertal Organizational Hormones." *Frontiers in Neuroendocrinology* 32:183–200.
Bering, Jesse. 2014. "This Queasy Love: How Having Frequent Diarrhea as a Child Shapes Your Adult Mate Choice." *Scientific American*, August 12. https://blogs.scientificamerican.com/bering-in-mind/this-queasy-love-how-having-frequent-diarrhea-as-a-child-shapes-your-adult-mate-choice/.
Berlant, Lauren, et al. 1994. "FORUM: On the Political Implications of Using the Term 'Queer,' as in 'Queer Politics,' 'Queer Studies,' and 'Queer Pedagogy.'" *Radical Teacher* 45:52–57.
Berman, Louis. 2003. *The Puzzle: Exploring the Evolutionary Puzzle of Male Homosexuality*. Wilmette, IL: Godot Press.
Berrios, German. 1996. *The History of Mental Symptoms: Descriptive Psychopathology since the 19th Century*. Cambridge: Cambridge University Press.
Bickerton, Derek. 1997. "Constructivism, Nativism, and Explanatory Adequacy." *Behavioral and Brain Sciences* 20:557–58.

Bieber, Irving. 1987. "On Arriving at the American Psychiatric Association Decision on Homosexuality." In *Scientific Controversies: Case-Studies in the Resolution and Closure in Science and Technology*, edited by H. Tristam Engelhardt Jr. and Arthur Caplan, 417–36. New York: Cambridge University Press.

Bieber, Irving, Harvey Dain, Paul Dince, Marvin Drellich, Henry Grand, Ralph Gundlach, Malvina Kremer, Alfred Rifkin, Cornelia Wilbur, and Toby Bieber. 1962. *Homosexuality: A Psychoanalytic Study of Male Homosexuals*. New York: Basic Books.

Blanchard, Ray. 2004. "Quantitative and Theoretical Analyses of the Relation between Older Brothers and Homosexuality in Men." *Journal of Theoretical Biology* 230:173–87.

———. 2018. "Fraternal Birth Order, Family Size, and Male Homosexuality: Meta-Analysis of Studies Spanning 25 Years." *Archives of Sexual Behavior* 47:1–15.

Blanchard, Ray, and Anthony F. Bogaert. 1996. "Homosexuality in Men and Number of Older Brothers." *American Journal of Psychiatry* 153 (1): 27–31.

Blank, Hanne. 2012. *Straight: The Surprisingly Short History of Heterosexuality*. Boston: Beacon Press.

Blankschaen, Kurt. 2019. "Rethinking Same-Sex Sex in Natural Law Theory." *Journal of Applied Philosophy* 37 (3): 428–45.

Block, Ned, and Gerald Dworkin. 1976. "IQ, Heritability and Inequality." In *The IQ Controversy*, edited by Ned Block and Gerald Dworkin, 410–540. New York: Pantheon Books.

Bobrow, David, and J. Michael Bailey. 2001. "Is Male Homosexuality Maintained via Kin Selection?" *Evolution and Human Behavior* 22:361–68.

Bogaert, Anthony. 2006. "Biological versus Nonbiological Older Brothers and Men's Sexual Orientation." *Proceedings of the National Academy of Sciences* 103:10771–74.

Bogaert, Anthony, and Scott Hershberger. 1999. "The Relation between Sexual Orientation and Penile Size." *Archives of Sexual Behavior* 28:213–21.

Bogaert, Anthony F., Malvina N. Skorska, Chao Wang, José Gabrie, Adam J. MacNeil, Mark R. Hoffarth, Doug P. VanderLaan, Kenneth J. Zucker, and Ray Blanchard. 2018. "Male Homosexuality and Maternal Immune Responsivity to the Y-Linked Protein NLGN4Y." *Proceedings of the National Academy of Sciences* 115 (2): 302–6.

Boorse, Christopher. 1975. "On the Distinction between Disease and Illness." *Philosophy and Public Affairs* 5:49–68.

———. 1977. "Health as a Theoretical Concept." *Philosophy of Science* 44:542–73.

———. 1997. "A Rebuttal on Health." In *What Is Disease?*, edited by J. M. Humber and R. F. Almeder, 1–134. Totowa, NJ: Humana Press.

———. 2010. "Disability and Medical Theory." In *Philosophical Reflections on Disability*, edited by D. Christopher Ralston and Justin Hubert Ho, 55–88. Dordrecht, Netherlands: Springer.

Borgerhoff-Mulder, Monique. 1988. "Behavioural Ecology in Traditional Societies." *Trends in Ecology & Evolution* 3:260–64.

Borris, Kenneth, and George Rousseau. 2008. *The Sciences of Homosexuality in Early Modern Europe*. London: Routledge.

Boswell, John. 1980. *Christianity, Social Tolerance, and Homosexuality: Gay People in Western Europe from the Beginning of the Christian Era to the Fourteenth Century*. Chicago: University of Chicago Press.

Bots, Jessica, Luc De Bruyn, Stefan Van Dongen, Roel Smolders, and Hans Van Gossum. 2009. "Female Polymorphism, Condition Differences, and Variation in

Male Harassment and Ambient Temperature." *Biological Journal of the Linnean Society* 97:545–54.

Bourget, D., and D. J. Chalmers. 2014. "What Do Philosophers Believe?" *Philosophical Studies* 170:465–500.

Boyd, Robert, and Peter Richerson. 1985. *Culture and the Evolutionary Process*. Chicago: University of Chicago Press.

Boysen, Guy, and David Vogel. 2007. "Biased Assimilation and Attitude Polarization in Response to Learning about Biological Explanations of Homosexuality." *Sex Roles* 57:755–62.

Bray, Alan. 2003. *The Friend*. Chicago: University of Chicago Press.

Brooks, Ross. 2009. "All Too Human: Responses to Same-Sex Copulation in the Common Cockchafer (Melolontha Melolontha [L.]), 1834–1900." *Archives of Natural History* 36 (1): 146–59.

———. 2010. "Transforming Sexuality: The Medical Sources of Karl Heinrich Ulrichs (1825–95) and the Origins of the Theory of Bisexuality." *Journal of the History of Medicine and Allied Sciences* 67:177–216.

———. 2021. "Darwin's Closet: The Queer Sides of *The Descent of Man* (1871)." *Zoological Journal of the Linnean Society* 191:323–46.

Buffon, Georges-Louis Leclerc de. 1954. *L'histoire naturelle, générale et particulière, tôme IV*. Edited by J. Piveteau. Paris: Presses Universitaires de France.

Bullough, Vern. 1998. "Alfred Kinsey and the Kinsey Report: Historical Overview and Lasting Contributions." *Journal of Sex Research* 35:127–31.

———. 2003. "The Contributions of John Money: A Personal View." *Journal of Sex Research* 40:230–36.

Button, Katherine S., John P. A. Ioannidis, Claire Mokrysz, Brian A. Nosek, Jonathan Flint, Emma S. J. Robinson, and Marcus R. Munafò. 2013. "Power Failure: Why Small Sample Size Undermines the Reliability of Neuroscience." *Nature Reviews Neuroscience* 14 (5): 365–76.

Byne, W., and E. Stein. 1997. "Ethical Implications of Scientific Research on the Causes of Sexual Orientation." *Health Care Analysis* 5:136–48.

Cadden, Joan. 2001. "'Nothing Natural Is Shameful': Vestiges of a Debate about Sex and Science in a Group of Late-Medieval Manuscripts." *Speculum* 76:66–89.

———. 2003. *Nothing Natural Is Shameful: Sodomy and Science in Late Medieval Europe*. Philadelphia: University of Pennsylvania Press.

Cammaerts, M. C. 2017. "Ants' Ability in Solving Simple Problems." *International Journal of Biology* 9:26–37.

Camperio-Ciani, Andrea, Umberto Battaglia, and Giovanni Zanzotto. 2015. "Human Homosexuality: A Paradigmatic Arena for Sexually Antagonistic Selection?" *Cold Spring Harbor Perspectives in Biology* 7 (4): a017657.

Camperio-Ciani, Andrea, Francesca Corna, and Claudio Capiluppi. 2004. "Evidence for Maternally Inherited Factors Favouring Male Homosexuality and Promoting Female Fecundity." *Proceedings of the Royal Society of London B* 271:2217–21.

Casper, Johann. 1863. *Klinische Novellen zur gerichtlichen Medicin: Nach eignen Erfahrungen*. Berlin: Hirschwald.

Cass, Vivienne. 1984a. "Homosexual Identity: A Concept in Need of Definition." *Journal of Homosexuality* 9:105–26.

———. 1984b. "Homosexual Identity Formation: Testing a Theoretical Model." *Journal of Sex Research* 20:143–67.

Cazzolla Gatti, Roberto. 2016. "Self-Consciousness: Beyond the Looking-Glass and What Dogs Found There." *Ethology Ecology & Evolution* 28:232–40.
Chang, Hasok. 2017. "Who Cares about the History of Science?" *Notes and Records: The Royal Society Journal of the History of Science* 71 (1): 91–107.
Chauncey, G. 1982. "From Sexual Inversion to Homosexuality: Medicine and the Changing Conceptualization of Female Deviance." *Salmagundi* 58/59:114–46.
———. 1997. *Gay New York: Gender, Urban Culture, and The Making of the Gay World, 1890–1940*. New York: Basic Books.
Cheng, S., and B. Powell. 2015. "Measurement, Methods, and Divergent Patterns: Reassessing the Effects of Same-Sex Parents." *Social Science Research* 52:615–26.
Chopra, Anuj. 2017. "Afghan Soldiers Are Using Boys as Sex Slaves, and the U.S. Is Looking the Other Way." *Washington Post*, July 19. https://www.washingtonpost.com/news/global-opinions/wp/2017/07/18/afghan-soldiers-are-using-boys-as-sex-slaves-and-the-u-s-is-looking-the-other-way/.
Christina, G. 2017. "Are We Having Sex Now or What?" In *The Philosophy of Sex: Contemporary Readings*, edited by Raja Halwani, Alan Soble, Sarah Hoffman, and Jacob M. Held, 31–38. Lanham, MD: Rowman & Littlefield.
Claudius Aelianus. 1666. *His Various History*. Translated by Thomas Stanley. London: Thomas Dring; Ann Arbor: Text Creation Partnership. https://quod.lib.umich.edu/e/eebo/A26482.0001.001/1:1?rgn=div1;view=fulltext.
Colledani, Daiana, and Andrea Camperio-Ciani. 2021. "A Worldwide Internet Study Based on Implicit Association Test Revealed a Higher Prevalence of Adult Males' Androphilia Than Ever Reported Before." *Journal of Sexual Medicine* 18 (1): 4–16.
Colmenares, Fernando, Heribert Hofer, and Marion East. 2000. "Greeting Ceremonies in Baboons and Hyenas." In *Natural Conflict Resolution*, edited by Filippo Aureli and Frans De Wall, 94–96. Berkeley: University of California Press.
Condon, Ed. 2019. "Study Finds No 'Gay Gene'—What That Means for Catholic Morality." *Catholic News Agency*, August 30. https://www.catholicnewsagency.com/news/study-finds-no-gay-gene---what-that-means-for-catholic-morality-44728.
Cooper, Rachel. 2002. "Disease." *Studies in History and Philosophy of Science Part C* 33:263–82.
———. 2005. *Classifying Madness: A Philosophical Examination of the Diagnostic and Statistical Manual of Mental Disorders*. Dordrecht, Netherlands: Springer.
Coria-Avila, Genaro A. 2012. "The Role of Conditioning on Heterosexual and Homosexual Partner Preferences in Rats." *Socioaffective Neuroscience & Psychology* 2 (1): 17340.
Corvino, John. 2013. *What's Wrong with Homosexuality?* Oxford: Oxford University Press.
Cottrell, Catherine, and Steven Neuberg. 2005. "Different Emotional Reactions to Different Groups: A Sociofunctional Threat-Based Approach to 'Prejudice.'" *Journal of Personality and Social Psychology* 88:770–89.
Coucke, Gijs. 2009. "Translation and Textual Criticism in the Middle Ages: Pierre of Abano's 'Expositio Problematum' (1310)." *Filologia Mediolatina* 16:187–213.
Crompton, Louis. 2003. *Homosexuality and Civilization*. Cambridge, MA: Harvard University Press.
Crozier, Ivan. 2008. "Havelock Ellis, Eugenicist." *Studies in History and Philosophy of Science Part C* 39:187–94.
Cryle, Peter, and Lisa Downing. 2009. "Feminine Sexual Pathologies." *Journal of the History of Sexuality* 18:1–7.

Darwin, Charles. 2008. *On the Origin of Species*. Oxford: Oxford University Press.
———. n.d. "Letter No. 4727." Darwin Correspondence Project.
Daston, Lorraine. 2010. "Human Nature Is a Garden." *Interdisciplinary Science Reviews* 35 (3–4): 215–30.
Davidson, Arnold. 1987. "How to Do the History of Psychoanalysis: A Reading of Freud's *Three Essays on the Theory of Sexuality*." *Critical Inquiry* 13:252–77.
———. 1991. "Closing up the Corpses: Diseases of Sexuality and the Emergence of the Psychiatric Style of Reasoning." In *Mind, Meaning and Method: Essays in Honor of Hilary Putnam*, edited by George Boolos, 295–325. Cambridge: Cambridge University Press.
———. 2001. *The Emergence of Sexuality: Historical Epistemology and the Formation of Concepts*. Cambridge, MA: Harvard University Press.
Davidson, Donald. 1982. "Rational Animals." *Dialectica* 36:317–28.
Davis, Katherine B. 1929. *Factors in the Sex Life of Twenty-Two Hundred Women*. New York: Harpers & Brothers.
Davis, Kingsley. 1937. "Kingsley Davis on Reproductive Institutions and the Pressure for Population." *Population and Development Review* 23:611–24.
De Block, Andreas, and Pieter R. Adriaens. 2004. "Darwinizing Sexual Ambivalence: A New Evolutionary Hypothesis of Male Homosexuality." *Philosophical Psychology* 17 (1): 59–76.
———. 2013. "Pathologizing Sexual Deviance: A History." *Journal of Sex Research* 50 (3–4): 276–98.
———. 2016. "Decreased Reproductive Success." In *Encyclopedia of Evolutionary Psychological Science*, edited by V. Weekes-Shackelford and T. K. Shackelford. New York: Springer.
De Block, Andreas, and Stijn Conix. Forthcoming. "Responsible Dissemination in Sexual Orientation Research: The Case of the AI 'Gaydar.'" *Philosophy of Science*.
De Block, Andreas, and Olivier Lemeire. 2015. "Philosophy and the Biology of Male Homosexuality." *Philosophy Compass* 10 (7): 479–88.
De Block, Andreas, and J. Sholl. 2020. "Harmless Dysfunctions and the Problem of Normal Variation." In *Defining Mental Disorders: Jerome Wakefield and His Critics*, edited by L. Faucher and D. Forest. Cambridge, MA: MIT Press.
De Boer, C. 1978. "The Polls: Attitudes toward Homosexuality." *Public Opinion Quarterly* 42:265–76.
De Bont, Raf. 2010. "Schizophrenia, Evolution and the Borders of Biology: On Huxley et al.'s 1964 Paper in Nature." *History of Psychiatry* 21 (2): 144–59.
De Cecco, John P., and John P. Elia, eds. 1993. *If You Seduce a Straight Person, Can You Make Them Gay? Issues in Biological Essentialism versus Social Constructionism in Gay and Lesbian Identities*. New York: Haworth Press.
Decker, Hannah. 2007. "How Kraepelinian Was Kraepelin? How Kraepelinian Were the Neo-Kraepelinians? From Emil Kraepelin to DSM-III." *History of Psychiatry* 18:337–60.
Dembroff, Robin. 2016. "What Is Sexual Orientation?" *Philosopher's Imprint* 16 (3): 1–27.
Despret, V. 2016. *What Would Animals Say If We Asked the Right Questions?* Minneapolis: University of Minnesota Press.
Detamore, Mathias. 2016. "Queer(y)Ing the Ethics of Research Methods: Toward a Politics of Intimacy in Researcher/Researched Relations." In *Queer Methods and*

Methodologies: Intersecting Queer Theories and Social Science Research, edited by Catherine J. Nash, 167–82. London: Routledge.

Diamond, Lisa. 2021. "The New Genetic Evidence on Same-Gender Sexuality: Implications for Sexual Fluidity and Multiple Forms of Sexual Diversity." *Journal of Sex Research* 58:1–20.

Díaz-León, Esa. Forthcoming. "Sexual Orientations: The Desire View." In *Feminist Philosophy of Mind*, edited by Keya Maitra and Jennifer McWeeny. Oxford: Oxford University Press.

Dickinson, R. L., and L. Beam. 1931. *A Thousand Marriages*. Baltimore, MD: Williams & Wilkins.

Doebner, E. 1850. "Über scheinbar abnorme Antennenform bei Melolontha vulgaris." *Entomologische Zeitung* 11:327–28.

Douglas, M. 1999. *Leviticus as Literature*. Oxford: Oxford University Press.

Downing, Lisa. 2010. "John Money's 'Normophilia': Diagnosing Sexual Normality in Late-Twentieth Century Anglo-American Sexology." *Psychology & Sexuality* 1:275–87.

Drescher, Jack. 2010. "Queer Diagnoses: Parallels and Contrasts in the History of Homosexuality, Gender Variance, and the Diagnostic and Statistical Manual." *Archives of Sexual Behavior* 39:427–60.

Drescher, Jack, and Joseph P. Merlino. 2007. *American Psychiatry and Homosexuality: An Oral History*. London: Routledge.

Drescher, Jack, Alan Schwartz, Flávio Casoy, Christopher A. McIntosh, Brian Hurley, Kenneth Ashley, Mary Barber, et al. 2016. "The Growing Regulation of Conversion Therapy." *Journal of Medical Regulation* 102 (2): 7–12.

Duyckaerts, François. 1966. *La formation du lien sexuel*. Bruxelles, Belgium: Dessart.

Dworek, Günter. 1990. "'Ist diese Krankheit heilbar?' Zwei Irrenärzte kommentieren Karl Heinrich Ulrichs." *Capri: Zeitschrift für schwule Geschichte* 8:42–56.

Eckert, Elke D., Thomas J. Bouchard, Joseph Bohlen, and Leonard L. Heston. 1986. "Homosexuality in Monozygotic Twins Reared Apart." *British Journal of Psychiatry* 148 (4): 421–25.

Ehrlich, P. R. 2000. *Human Natures: Genes, Cultures, and the Human Prospect*. Washington, DC: Island Press.

Eigen, Joel. 1995. *Witnessing Insanity: Madness and Mad-Doctors in the English Court*. New Haven, CT: Yale University Press.

Ellenberger, Henri. 1970. *The Discovery of the Unconscious: The History and Evolution of Dynamic Psychiatry*. New York: Basic Books.

Ellis, Christopher, and James Stimson. 2012. *Ideology in America*. Cambridge: Cambridge University Press.

Ellis, Havelock. 1900. *Studies in the Psychology of Sex*. London: London University Press.

Ellis, Havelock, and John Symonds. 2008. *Sexual Inversion: A Critical Edition*. Edited by Ivan Crozier. Basingstoke, UK: Palgrave MacMillan.

Endler, John A. 1986. *Natural Selection in the Wild*. Princeton, NJ: Princeton University Press.

Ennis, Dawn. 2019. "The 'Gay Gene' Is A Myth But Being Gay Is 'Natural,' Say Scientists." *Forbes*, August 30. https://www.forbes.com/sites/dawnstaceyennis/2019/08/30/the-gay-gene-is-a-myth-but-being-gay-is-natural-say-scientists/#57141ceb7fa7.

Ereshefsky, Marc. 2009. "Defining 'Health' and 'Disease.'" *Studies in History and Philosophy of Science Part C* 40 (3): 221–27.

Eriksson, Kimmo, and Pontus Strimling. 2015. "Group Differences in Broadness of Values May Drive Dynamics of Public Opinion on Moral Issues." *Mathematical Social Sciences* 77 (September): 1–8.

Fasoli, F., and A. Maass. 2018. "Voice and Prejudice: The Social Costs of Auditory Gaydar." *Atlantic Journal of Communication* 26:98–110.

Fausto-Sterling, Anne. 1995. "Animal Models for the Development of Human Sexuality: A Critical Evaluation." *Journal of Homosexuality* 28:217–35.

———. 2000. *Sexing the Body: Gender Politics and the Construction of Sexuality*. New York: Basic Books.

Fedoroff, J. Paul. 2009. "The Paraphilias." In *The New Oxford Textbook of Psychiatry*, 2nd ed., edited by Michael Gelder, Nancy Andreasen, Juan Lopez-Ibor, and John Geddes, 832–42. Oxford: Oxford University Press.

Féray, Jean-Claude. 2004. *Grecques, les moeurs du hanneton? Histoire du mot pédérastie et de ses dérivés en langue française*. Paris: Quintes-Feuilles.

Féray, Jean-Claude, Manfred Herzer, and Glen W. Peppel. 1990. "Homosexual Studies and Politics in the 19th Century: Karl Maria Kertbeny." *Journal of Homosexuality* 19 (1): 23–48.

Féré, Charles. 1896. *La descendance d'un inverti: Contribution à l'hygiène de l'inversion sexuelle*. Paris: Imprimerie Charles Schlaeber.

———. 1897. "Les perversions sexuelles chez les animaux." *Revue philosophique* 43:494–503.

———. 1899. *L'instinct sexuel: Evolution et dissolution*. Paris: Félix Alcan.

Fine, Cordelia. 2010. *Delusions of Gender: How Our Minds, Society, and Neurosexism Create Difference*. New York: W. W. Norton.

First, Michael. 2010. "DSM-5 Proposals for Paraphilias: Suggestions for Reducing False Positives Related to Use of Behavioral Manifestations." *Archives of Sexual Behavior* 39:1239–44.

First, Michael, and Allen Frances. 2008. "Issues for DSM-V: Unintended Consequences of Small Changes: The Case of Paraphilias." *American Journal of Psychiatry* 165: 1240–41.

Fisher, J. A. 1996. "The Myth of Anthropomorphism." In *Readings in Animal Cognition*, edited by M. Bekoff and D. Jamieson, 3–16. Cambridge, MA: MIT Press.

Fleischman, Diana, Daniel Fessler, and A. Evelyn Cholakians. 2014. "Testing the Affiliation Hypothesis of Homoerotic Motivation in Humans: The Effects of Progesterone and Priming." *Archives of Sexual Behavior* 44 (5):1395–1404.

Ford, Clellan, and Frank Beach. 1951. *Patterns of Sexual Behavior*. New York: Harpers & Brothers.

Foucault, Michel. 1978. *The History of Sexuality*. Vol. 1, *An Introduction*. Translated by Robert Hurley. New York: Pantheon Books.

Fox, Elizabeth A. 2001. "Homosexual Behavior in Wild Sumatran Orangutans (Pongo Pygmaeus Abelii)." *American Journal of Primatology* 55 (3): 177–81.

Freud, Sigmund. 1955. "The Psychogenesis of a Case of Homosexuality in a Woman." In *The Standard Edition of the Complete Psychological Works of Sigmund Freud*, edited by J. Strachey, 18:145–72. London: Hogarth Press.

———. 1960a. "Anonymous (Letter to an American Mother)." In *The Letters of Sigmund Freud*, edited by E. Freud, 423–24. London: Hogarth Press.

———. 1960b. "Three Essays on the Theory of Sexuality." In *The Standard Edition of the Complete Psychological Works of Sigmund Freud*, edited by J. Strachey, 7:123–246. London: Hogarth Press.

———. 1964a. "Analysis Terminable and Interminable." In *The Standard Edition of the Complete Psychological Works of Sigmund Freud*, edited by J. Strachey, 23:209–53. London: Hogarth Press.

———. 1964b. "Fetishism." In *The Standard Edition of the Complete Psychological Works of Sigmund Freud*, edited by J. Strachey, 21:153–61. London: Hogarth Press.

Friedman, Richard, and Jennifer Downey. 1998. "Psychoanalysis and the Model of Homosexuality as Psychopathology: A Historical Overview." *American Journal of Psychiatry* 58:249–70.

Frost-Arnold, Greg. 2011. "From the Pessimistic Induction to Semantic Antirealism." *Philosophy of Science* 78:1131–42.

Fuller Torrey, Edward. 1974. *The Death of Psychiatry*. Radnor, PA: Chilton Book Company.

Gadeau de Kerville, Henri. 1896a. *Observations relatives à ma note intitulée "Perversion sexuelle chez les coléoptères mâles."* Rouen, France: Lecerf.

———. 1896b. "Perversion sexuelle chez les coléoptères mâles." *Bulletin de la Société Entomologique de France* 4:85–87.

Gangestad, Steven W., J. Michael Bailey, and Nicholas G. Martin. 2000. "Taxometric Analyses of Sexual Orientation and Gender Identity." *Journal of Personality and Social Psychology* 78:1109.

Ganna, Andrea, Karin J. H. Verweij, Michel G. Nivard, Robert Maier, Robbee Wedow, Alexander S. Busch, Abdel Abdellaoui, et al. 2019. "Large-Scale GWAS Reveals Insights into the Genetic Architecture of Same-Sex Sexual Behavior." *Science* 365 (6456): eaat7693.

Gavrilets, Sergey, Urban Friberg, and William R. Rice. 2018. "Understanding Homosexuality: Moving On from Patterns to Mechanisms." *Archives of Sexual Behavior* 47 (1): 27–31.

Gelman, Andrew, Greggor Mattson, and Daniel Simpson. 2018. "Gaydar and the Fallacy of Decontextualized Measurement." *Sociological Science* 5:270–80.

Gelman, Susan A. 2003. *The Essential Child: Origins of Essentialism in Everyday Thought*. New York: Oxford University Press.

———. 2004. "Psychological Essentialism in Children." *Trends in Cognitive Sciences* 8:404–9.

Gelman, Susan A., and Henry Wellman. 1991. "Insides and Essences: Early Understandings of the Non-Obvious." *Cognition* 38:213–44.

Gerard, Kent, and Gert Hekma. 1989. *The Pursuit of Sodomy: Male Homosexuality in Renaissance and Enlightenment Europe*. New York: Harrington Park Press.

Gert, Bernard, and Charles Culver. 2009. "Sex, Immorality, and Mental Disorders." *Journal of Medicine and Philosophy* 34:487–95.

Ghiselin, Michael. 2005. "Homology as a Relation of Correspondence between Parts of Individuals." *Theory in Biosciences* 124:91–103.

Gilbert, Arthur, and Michael Barkun. 1981. "Disaster and Sexuality." *Journal of Sex Research* 17:288–99.

GLAAD. 2016. "GLAAD Media Reference Guide, 10th Edition." https://www.glaad.org/reference.

Godfrey-Smith, P. 2001. "Three Kinds of Adaptationism." In *Adaptationism and Optimality*, edited by S. H. Orzack and E. Sober, 335–57. New York: Cambridge University Press.

Gold, Ronald. 1973. "Stop It, You're Making Me Sick!" *American Journal of Psychiatry* 130:1211–12.
Goldberg, D. S. 2014. "Fatness, Medicalization, and Stigma: On the Need to Do Better." *Narrative Inquiry in Bioethics* 4:117–23.
Goldman, Alan. 1977. "Plain Sex." *Philosophy and Public Affairs* 6:267–87.
Goodhart, Charles. 1957. "The Future of Human Fertility." *New Scientist*, December 12.
Gooren, Louis. 2006. "The Biology of Human Psychosexual Differentiation." *Hormones and Behavior* 50:589–601.
Gowaty, Patricia Adair. 1982. "Sexual Terms in Sociobiology: Emotionally Evocative and, Paradoxically, Jargon." *Animal Behaviour* 30 (2): 630–31.
Gray, Robert. 1997. "Sex and Sexual Perversion." In *The Philosophy of Sex*, 3rd ed., edited by Alan Soble, 57–66. Lanham, MD: Rowman & Littlefield.
Greenberg, D. F. 1990. *The Construction of Homosexuality*. Chicago: University of Chicago Press.
Grene, Marjorie, and David Depew. 2004. *The Philosophy of Biology: An Episodic History*. Cambridge: Cambridge University Press.
Griffiths, Paul. 2002. "What Is Innateness?" *The Monist* 85:70–85.
Griffiths, Paul, Edouard Machery, and Stefan Linquist. 2009. "The Vernacular Concept of Innateness." *Mind & Language* 24 (5): 605–30.
Grinde, Bjorn. 2022. "The Contribution of Sex to Quality of Life in Modern Societies." *Applied Research in Quality of Life* 17:449–65.
Grob, Gerald. 1991. "Origins of DSM-I: A Study in Appearance and Reality." *American Journal of Psychiatry* 148:421–31.
Groneman, Carol. 1994. "Nymphomania: The Historical Construction of Female Sexuality." *Signs* 19:337–67.
Gutmann, Philipp. 2006. "On the Way to a 'Scientia Sexualis': 'On the Relation of the Sexual System to the Psyche in General and to Cretinism in Particular' (1826) by Joseph Häussler." *History of Psychiatry* 17:45–53.
Halperin, David M. 1990. *One Hundred Years of Homosexuality and Other Essays in Greek Love*. New York: Routledge.
———. 2002. *How to Do the History of Homosexuality*. Chicago: University of Chicago Press.
Hamer, D., S. Hu, V. Magnuson, N. Hu, and A. Pattatucci. 1993. "A Linkage between DNA Markers on the X Chromosome and Male Sexual Orientation." *Science* 261 (5119): 321–27.
Hamer, Dean H., and Peter Copeland. 1994. *The Science of Desire: The Search for the Gay Gene and the Biology of Behavior*. New York: Simon & Schuster.
Hare, Edward. 1962. "Masturbatory Insanity: The History of an Idea." *Journal of Mental Science* 108:2–25.
Harrold, Max. 1999. "Creature Comforts." *Advocate*, February 16.
Hartenstein, Christiana C., and John C. Gonsiorek. 2015. "Situational Homosexuality." In *The International Encyclopedia of Human Sexuality*, edited by Anne Bolin and Patricia Whelehan, 1115–1354. Oxford: John Wiley & Sons.
Haslam, Nick, and Sheri Levy. 2006. "Essentialist Beliefs about Homosexuality: Structure and Implications for Prejudice." *Personality and Social Psychology Bulletin* 32:471–85.
Haslanger, Sally. 2000. "Gender and Race: (What) Are They? (What) Do We Want Them to Be?" *Noûs* 34:31–55.
———. 2003. "Social Construction: The 'Debunking' Project." In *Socializing Metaphysics*, edited by Frederick F. Schmitt, 301–25. Oxford: Rowman & Littlefield.

Hauser, R. I. 1992. "Sexuality, Neurasthenia and the Law: Richard von Krafft-Ebing (1840–1902)." PhD diss., University of London.
Hayenhjelm, M., and J. Wolff. 2012. "The Moral Problem of Risk Impositions: A Survey of the Literature." *European Journal of Philosophy* 20:E26–51.
Hegarty, P., and F. Pratto. 2001. "The Effects of Social Category Norms and Stereotypes on Explanations for Intergroup Differences." *Journal of Personality and Social Psychology* 80 (5): 723.
Hegarty, Peter, and Anne Golden. 2008. "Attributions about the Controllability of Stigmatized Traits: Antecedents or Justifications of Prejudice?" *Journal of Applied Social Psychology* 38:1023–44.
Hekma, G. 1991. "Homosexual Behavior in the Nineteenth-Century Dutch Army." *Journal of the History of Sexuality* 2 (2): 266–88.
Henrich, Joseph. 2017. *The Secret of Our Success: How Culture Is Driving Human Evolution, Domesticating Our Species and Making Us Smarter*. Princeton, NJ: Princeton University Press.
———. 2020. *The WEIRDest People in the World*. New York: Farrar, Straus and Giroux.
Hensley, Christopher, and Richard Tewksbury. 2002. "Inmate-to-Inmate Prison Sexuality: A Review of Empirical Studies." *Trauma, Violence & Abuse* 3 (3): 226–43.
Herdt, Gilbert H. 1994. *Guardians of the Flutes: Idioms of Masculinity*. Chicago: University of Chicago Press.
———. 1999. *Sambia Sexual Culture: Essays from the Field*. Chicago: University of Chicago Press.
———. 2019. "Intimate Consumption and New Sexual Subjects among the Sambia of Papua New Guinea." *Oceania* 89 (1): 36–67.
Herdt, Gilbert H., and Martha McClintock. 2000. "The Magical Age of 10." *Archives of Sexual Behavior* 29 (6): 587–606.
Herek, Gregory. 1994. "Assessing Heterosexuals' Attitudes toward Lesbians and Gay Men: A Review of Empirical Research with the ATLG Scale." In *Lesbian and Gay Psychology: Theory, Research, and Clinical Applications*, edited by Beverly Greene and Gregory Herek, 206–28. Thousand Oaks, CA: Sage Publications.
———. 2004. "Beyond 'Homophobia': Thinking about Sexual Prejudice and Stigma in the Twenty-First Century." *Sexuality Research and Social Policy* 1:6–24.
Herzer, Manfred. 1986. "Kertbeny and the Nameless Love." *Journal of Homosexuality* 12 (1): 1–26.
Hesslow, Germund. 1993. "Do We Need a Concept of Disease?" *Theoretical Medicine* 14:1–14.
Hewitt, Christopher. 1995. "The Socioeconomic Position of Gay Men: A Review of the Evidence." *American Journal of Economics and Sociology* 54 (4): 461–79.
Heyes, Cecilia. 2016. "Who Knows? Metacognitive Social Learning Strategies." *Trends in Cognitive Sciences* 20 (3): 204–13.
Hilker, Rikke, Dorte Helenius, Birgitte Fagerlund, Axel Skytthe, Kaare Christensen, Thomas M. Werge, Merete Nordentoft, and Birte Glenthøj. 2018. "Heritability of Schizophrenia and Schizophrenia Spectrum Based on the Nationwide Danish Twin Register." *Biological Psychiatry* 83 (6): 492–98.
Hill, Darryl. 2005. "Sexuality and Gender in Hirschfeld's *Die Transvestiten*: A Case of the 'Elusive Evidence of the Ordinary.'" *Journal of the History of Sexuality* 14: 316–32.
Hirschfeld, Magnus. 1914. *Die Homosexualität des Mannes und des Weibes*. Berlin: Louis Marcus Verlagsbuchhandlung.

———. 1952. *Sexual Anomalies and Sexual Perversions*. London: Encyclopaedic Press.
Hofman, Elwin. 2021. *Trials of the Self: Murder, Mayhem, and the Remaking of the Mind, 1750–1830*. Manchester: Manchester University Press.
Hopcroft, Rosemary L. 2006. "Sex, Status, and Reproductive Success in the Contemporary United States." *Evolution and Human Behavior* 27 (2): 104–20.
Horwitz, Allan V., and Jerome C. Wakefield. 2007. *The Loss of Sadness: How Psychiatry Transformed Normal Sorrow into Depressive Disorder*. New York: Oxford University Press.
Howard, Judith A. 2000. "Social Psychology of Identities." *Annual Review of Sociology* 26 (1): 367–93.
Hull, David. 1998. "On Human Nature." In *The Philosophy of Biology*, edited by David Hull and Michael Ruse, 383–98. New York: Oxford University Press.
Hutchinson, G. E. 1953. *The Itinerant Ivory Tower: Scientific and Literary Essays*. New Haven, CT: Yale University Press.
———. 1957. "Marginalia." *American Scientist* 45:88–96.
———. 1959. "A Speculative Consideration of Certain Possible Forms of Sexual Selection in Man." *American Naturalist* 93 (869): 81–91.
———. 1979. *The Kindly Fruits of the Earth: Recollections of an Embryo Ecologist*. New Haven, CT: Yale University Press.
Huxley, Julian, Ernst Mayr, Humphry Osmond, and Abram Hoffer. 1964. "Schizophrenia as a Genetic Morphism." *Nature* 204 (4955): 220–21.
Iemmola, Francesca, and Andrea Camperio-Ciani. 2009. "New Evidence of Genetic Factors Influencing Sexual Orientation in Men: Female Fecundity Increase in the Maternal Line." *Archives of Sexual Behavior* 38 (3): 393–99.
Inglehart, Ronald, and Christian Welzel. 2005. *Modernization, Cultural Change, and Democracy: The Human Development Sequence*. New York: Cambridge University Press.
Institut für Sexualwissenschaft, Dr. Magnus Hirschfeld-Stiftung. 1924. *Unsere Arbeit*. Berlin: NW 40.
Isler, Karin, and Carel P. Van Schaik. 2009. "Why Are There So Few Smart Mammals (but So Many Smart Birds)?" *Biology Letters* 5 (1): 125–29.
Isomura, S., and M. Mizogami. 1992. "The Low Rate of HIV Infection in Japanese Homosexual and Bisexual Men: An Analysis of HIV Seroprevalence and Behavioural Risk Factors." *AIDS* 6:501–3.
Jacobs, Jürgen. 1981. "How Heritable Is Innate Behaviour?" *Zeitschrift Für Tierpsychologie* 55 (1): 1–18.
James, William. 1887. "What Is an Instinct?" *Scribner's Magazine* 1:355–65.
Jannini, Emmanuele A., Ray Blanchard, Andreas Camperio-Ciani, and John Bancroft. 2010. "Controversies in Sexual Medicine: Male Homosexuality: Nature or Culture?" *Journal of Sexual Medicine* 7 (10): 3245–53.
Jayaratne, Toby Epstein, Oscar Ybarra, Jane P. Sheldon, Tony N. Brown, Merle Feldbaum, Carla Pfeffer, and Elizabeth M. Petty. 2006. "White Americans' Genetic Lay Theories of Race Differences and Sexual Orientation: Their Relationship with Prejudice toward Blacks, and Gay Men and Lesbians." *Group Processes & Intergroup Relations* 9 (1).
Jordan, Mark. 1997. *The Invention of Sodomy in Christian Theology*. Chicago: University of Chicago Press.
Kahan, Benjamin. 2021. "The Unexpected American Origins of Sexology and Sexual Science: Elizabeth Osgood Goodrich Willard, Orson Squire Fowler, and the Scientification of Sex." *History of the Human Sciences* 34 (1): 71–88.

Kallmann, Franz J. 1938. *The Genetics of Schizophrenia*. New York: J. J. Augustin.
———. 1946. "The Genetic Theory of Schizophrenia." *American Journal of Psychiatry* 103:309–22.
———. 1952a. "Comparative Twin Study on the Genetic Aspects of Male Homosexuality." *Journal of Nervous and Mental Disease* 115 (1): 283–98.
———. 1952b. "Twin and Sibship Study of Overt Male Homosexuality." *American Journal of Human Genetics* 4 (2): 136–46.
———. 1953. *Heredity in Health and Mental Disorder; Principles of Psychiatric Genetics in the Light of Comparative Twin Studies*. New York: W. W. Norton.
Kamieniak, Jean-Pierre. 2003. "La construction d'un objet psychopathologique: La perversion sexuelle au XIXe siècle." *Revue française de psychanalyse* 67:249–62.
Karsch, Ferdinand. 1900. "Päderastie und Tribadie bei den Tieren auf Grund der Literatur." *Jahrbuch für sexuelle Zwischenstufen* 2:126–60.
Kelch, August. 1834. "Beobachtung über die Bastardbegattung zwischen Melolontha vulgaris und Melolontha hippocastani." *Isis von Oken* 7:737–38.
Keller, Johannes. 2005. "In Genes We Trust: The Biological Component of Psychological Essentialism and Its Relationship to Mechanisms of Motivated Social Cognition." *Journal of Personality and Social Psychology* 88:686–702.
Kelley, Ken. 1978. "Playboy Interview: Anita Bryant." *Playboy Magazine*, May.
King, Michael, John Green, David P. J. Osborn, Jamie Arkell, Jacqueline Hetherton, and Elizabeth Pereira. 2005. "Family Size in White Gay and Heterosexual Men." *Archives of Sexual Behavior* 34 (1): 117–22.
Kinsey, Alfred, Clyde Martin, and Wardell Pomeroy. 1948. *Sexual Behavior in the Human Male*. Philadelphia: W. B. Saunders.
Kinsey, Alfred, Clyde Martin, Wardell Pomeroy, and Paul Gebhard. 1953. *Sexual Behavior in the Human Female*. Philadelphia: W. B. Saunders.
Kirk, Stuart, and Herb Kutchins. 1992. *The Selling of the DSM: The Rhetoric of Science in Psychiatry*. New York: Aldine De Gruyter.
Kirkpatrick, R. Craig. 2000. "The Evolution of Human Homosexual Behavior." *Current Anthropology* 41 (3): 385–413.
Kitcher, Philip. 2001. *Science, Truth, and Democracy*. New York: Oxford University Press.
Knauft, Bruce. 1985. *Good Company and Violence: Sorcery and Social Action in a Lowland New Guinea Society*. Berkeley: University of California Press.
Knight, Raymond A., and Robert A. Prentky. 1990. "Classifying Sexual Offenders: The Development and Corroboration of Taxonomic Models." In *Handbook of Sexual Assault*, edited by William Lamont Marshall, D. R. Laws, and Howard E. Barbaree. New York: Springer.
Kosinski, M., D. Stillwell, and T. Graepel. 2013. "Private Traits and Attributes Are Predictable from Digital Records of Human Behavior." *Proceedings of the National Academy of Sciences* 110:5802–5.
Kotrschal, Kurt, Josef Hemetsberger, and Brigitte Weiß. 2006. "Making the Best of a Bad Situation: Homosociality in Male Greylag Geese." In *Homosexual Behaviour in Animals: An Evolutionary Perspective*, edited by Volker Sommer and Paul L. Vasey, 45–76. Cambridge: Cambridge University Press.
Kourany, Janet A. 2016. "Should Some Knowledge Be Forbidden? The Case of Cognitive Differences Research." *Philosophy of Science* 83 (5): 779–90.
Krafft-Ebing, Richard von. 1886. *Psychopathia sexualis: Eine klinisch-forensische Studie*. Stuttgart, Germany: Ferdinand Enke.

Kuefler, M. 2007. "The Marriage Revolution in Late Antiquity: The Theodosian Code and Later Roman Marriage Law." *Journal of Family History* 32:343–70.

Kukla, Quill R. 2019. "Infertility, Epistemic Risk, and Disease Definitions." *Synthese* 196 (11): 4409–28.

Laboulbène, Alexandre. 1859. "Examen anatomique de deux Melolontha vulgaris trouvés accouplés et paraissant du sexe mâle." *Annales de la Société Entomologique de France* 3:567–70.

Lacassagne, Alexandre. 1882. *De la criminalité chez les animaux*. Lyon, France: Imprimerie de L. Bourgeon.

Lakatos, Imre. 1970. "History of Science and Its Rational Reconstructions." In *PSA 1970*, edited by Roger C. Buck and Robert S. Cohen, 8:91–136. Dordrecht, Netherlands: Springer.

Långström, Niklas, Qazi Rahman, Eva Carlström, and Paul Lichtenstein. 2010. "Genetic and Environmental Effects on Same-Sex Sexual Behavior: A Population Study of Twins in Sweden." *Archives of Sexual Behavior* 39 (1): 75–80.

Laplane, Lucie, Paolo Mantovani, Ralph Adolphs, Hasok Chang, Alberto Mantovani, Margaret McFall-Ngai, Carlo Rovelli, Elliott Sober, and Thomas Pradeu. 2019. "Opinion: Why Science Needs Philosophy." *Proceedings of the National Academy of Sciences* 116 (10): 3948–52.

Laqueur, Thomas. 1992. *Making Sex: Body and Gender from the Greeks to Freud*. Cambridge, MA: Harvard University Press.

Lautmann, Rudiger. 1980. "The Pink Triangle: The Persecution of Homosexual Males in Concentration Camps in Nazi Germany." *Journal of Homosexuality* 6 (1–2): 141–60.

Leuner, J. 2019. "A Replication Study: Machine Learning Models Are Capable of Predicting Sexual Orientation from Facial Images." *ArXiv Preprint*:1902.10739.

LeVay, Simon. 1996. *Queer Science: The Use and Abuse of Research into Homosexuality*. Cambridge, MA: MIT Press.

———. 2011. *Gay, Straight, and the Reason Why: The Science of Sexual Orientation*. New York: Oxford University Press.

Levin, Michael. 1984. "Why Homosexuality Is Abnormal." *The Monist* 67:251–83.

Lewis, Paul. 2018. "'I Was Shocked It Was So Easy': Meet the Professor Who Says Facial Recognition Can Tell If You're Gay." *Guardian*, July 7. https://www.theguardian.com/technology/2018/jul/07/artificial-intelligence-can-tell-your-sexuality-politics-surveillance-paul-lewis.

Lewontin, Richard Charles. 1982. *Human Diversity*. New York: Scientific American Library.

Lewontin, Richard Charles, S. Rose, and L. J. Kamin. 1984. *Not in Our Genes*. New York: Pantheon Books.

Lloyd, Elisabeth. 2005. *The Case of the Female Orgasm: Bias in the Science of Evolution*. Cambridge, MA: Harvard University Press.

Löfström, J. 1997. "The Birth of the Queen/the Modern Homosexual: Historical Explanations Revisited." *Sociological Review* 45:24–41.

Lombroso, Cesare, and Guglielmo Ferrero. 1999. *The Female Offender*. Buffalo, NY: Fred B. Rothman Publications.

MacIntyre, Ferren, and Kenneth W. Estep. 1993. "Sperm Competition and the Persistence of Genes for Male Homosexuality." *Biosystems* 31 (2–3): 223–33.

Maitra, Poulami, Melissa Caughey, Laura Robinson, Payal C. Desai, Susan Jones, Mehdi Nouraie, Mark T. Gladwin, Alan Hinderliter, Jianwen Cai, and Kenneth I. Ataga.

2017. "Risk Factors for Mortality in Adult Patients with Sickle Cell Disease: A Meta-Analysis of Studies in North America and Europe." *Haematologica* 102 (4): 626–36.

Mallon, Ron. 2016. *The Construction of Human Kinds.* Oxford: Oxford University Press.

Mallon, Ron, and Stephen P. Stich. 2000. "The Odd Couple: The Compatibility of Social Construction and Evolutionary Psychology." *Philosophy of Science* 67:133–54.

Malm, Sara. 2017. "He's My Mane Man: Gay Lions Are Spotted Putting on Very Rare Public Display of Affection in Kenya." *Daily Mail*, November 1.

Mameli, Matteo. 2008. "On Innateness: The Clutter Hypothesis and the Cluster Hypothesis." *Journal of Philosophy* 105:719–36.

Mameli, Matteo, and Patrick Bateson. 2011. "An Evaluation of the Concept of Innateness." *Philosophical Transactions of the Royal Society B* 366 (1563): 436–43.

Mann, Janet. 2006. "Establishing Trust: Socio-Sexual Behaviour and the Development of Male-Male Bonds among Indian Ocean Bottlenose Dolphins." In *Homosexual Behaviour in Animals: An Evolutionary Perspective*, edited by Volker Sommer and Paul L. Vasey, 107–30. Cambridge: Cambridge University Press.

Markie, Peter J., and Timothy Patrick. 1990. "De Re Desire." *Australasian Journal of Philosophy* 68 (4): 432–47.

Marmor, Judd. 1973. "Homosexuality and Cultural Value Systems." *American Journal of Psychiatry* 130:1208–9.

Marques, Teresa. 2017. "The Relevance of Causal Social Construction." *Journal of Social Ontology* 3:1–25.

Marson, Lesley, and Ursula Wesselmann. 2017. "Female Sexual Function." In *Principles of Gender-Specific Medicine*, edited by Marianne J. Legato, 45–60. London: Elsevier.

Matte, Nicholas. 2005. "International Sexual Reform and Sexology in Europe, 1893–1933." *Canadian Bulletin of Medical History* 22:253–70.

Matthewson, John, and Paul E. Griffiths. 2017. "Biological Criteria of Disease: Four Ways of Going Wrong." *Journal of Medicine and Philosophy* 42 (4): 447–66.

Mattson, Greggor. 2017. "Dating Profiles Are Like Gay Bars: Peer Review, Ethics and LGBTQ Big Data." *Who Needs Gay Bars?* (blog), September 13. https://greggor mattson.com/2017/09/13/peer-review-ethics-and-lgbtq-big-data/.

Maxmen, J. 1985. *The New Psychiatry*. New York: Morrow.

McCarthy, Margaret M., Lindsay A. Pickett, Jonathan W. VanRyzin, and Katherine E. Kight. 2015. "Surprising Origins of Sex Differences in the Brain." *Hormones and Behavior* 76 (November): 3–10.

McFadden, D. 2002. "Masculinization Effects in the Auditory System." *Archives of Sexual Behavior* 31:99–111.

McHugh, Susan. 2011. *Animal Stories: Narrating across Species Lines*. Minneapolis: University of Minnesota Press.

McKnight, Jim. 1997. *Straight Science: Homosexuality, Evolution and Adaptation*. London: Routledge.

Medin, Douglas, and Andrew Ortony. 1989. "Comments on Part I: Psychological Essentialism." In *Similarity and Analogical Reasoning*, edited by Stella Vosniadou and Andrew Ortony, 179–96. Cambridge: Cambridge University Press.

Mehta, S., and A. Farina. 1997. "Is Being 'Sick' Really Better? Effect of the Disease View of Mental Disorder on Stigma." *Journal of Social and Clinical Psychology* 16:405–19.

Meisenberg, Gerhard. 2010. "The Reproduction of Intelligence." *Intelligence* 38 (2): 220–30.

Metzl, Jonathan. 2004. "Voyeur Nation? Changing Definitions of Voyeurism, 1950–2004." *Harvard Review of Psychiatry* 12:127–31.

Meyenburg, Bernd, and Volkmar Sigusch. 1977. "Sexology in West Germany." *Journal of Sex Research* 13:197–209.

Meyerowitz, Joanne. 2001. "Sex Research at the Borders of Gender: Transvestites, Transsexuals, and Alfred C. Kinsey." *Bulletin of the History of Medicine* 75:72–90.

Mildenberger, Florian. 2007. "Kraepelin and the 'Urnings': Male Homosexuality in Psychiatric Discourse." *History of Psychiatry* 18 (3): 321–35.

Mlinarić, Ana, Martina Horvat, and Vesna Šupak Smolčić. 2017. "Dealing with the Positive Publication Bias: Why You Should Really Publish Your Negative Results." *Biochemia Medica* 27 (3): 030201.

Moll, Albert. 1898. *Untersuchungen über die Libido sexualis*. Berlin: Fischer's Medicin Buchhandlung.

Money, John. 2003. "History, Causality, and Sexology." *Journal of Sex Research* 40:237–39.

Moon, Jordan W. 2020. "Why Are World Religions So Concerned with Sexual Behavior?" *Current Opinion in Psychology* 40 (August): 15–19.

Moran, P. A. P. 1972. "Familial Effects in Schizophrenia and Homosexuality." *Australian & New Zealand Journal of Psychiatry* 6 (2): 116–19.

Morton, T. A., M. J. Hornsey, and T. Postmes. 2009. "Shifting Ground: The Variable Use of Essentialism in Contexts of Inclusion and Exclusion." *British Journal of Social Psychology* 48 (1): 35–59.

Moser, Charles, and Peggy Kleinplatz. 2005. "DSM-IV-TR and the Paraphilias: An Argument for Removal." *Journal of Psychology & Human Sexuality* 17:91–109.

Murphy, Timothy. 1997. *Gay Science: The Ethics of Sexual Orientation Research*. New York: Columbia University Press.

Murray, Stephen. 2000. *Homosexualities*. Chicago: University of Chicago Press.

Muscarella, Frank. 2000. "The Evolution of Homoerotic Behavior in Humans." *Journal of Homosexuality* 40:51–77.

Musick, M. A. 2014. "A Review of Methodological and Ethical Issues Surrounding the New Family Structures Study." Unpublished report.

Mustanski, Brian S., Meredith L. Chivers, and J. Michael Bailey. 2002. "A Critical Review of Recent Biological Research on Human Sexual Orientation." *Annual Review of Sex Research* 13:89–140.

Mustanski, Brian S., Michael G. DuPree, Caroline M. Nievergelt, Sven Bocklandt, Nicholas J. Schork, and Dean H. Hamer. 2005. "A Genomewide Scan of Male Sexual Orientation." *Human Genetics* 116 (4): 272–78.

Nachshoni, Kobi. 2016. "Ramat Gan's Chief Rabbi: Gays and Lesbians Are Disabled and Predatory." *Ynetnews*, July 17. https://www.ynetnews.com/articles/0,7340,L-4829648,00.html.

Nagel, Thomas. 1969. "Sexual Perversion." *Journal of Philosophy* 66 (1): 5–17.

National Committee for Mental Hygiene. 1918. *Statistical Manual for the Use of Hospitals for Mental Diseases*. New York: National Committee for Mental Hygiene.

National Research Council. 2004. *Biotechnology Research in an Age of Terrorism*. Washington, DC: National Academies Press.

Nelson, Nicole C. 2018. *Model Behavior: Animal Experiments, Complexity, and the Genetics of Psychiatric Disorders*. Chicago: University of Chicago Press.

Nettle, Daniel, and Helen Clegg. 2005. "Schizotypy, Creativity and Mating Success in Humans." *Proceedings of the Royal Society of London B* 273:611–15.

Newson, Lesley, Stephen Lea, Tom Postmes, and Paul Webley. 2005. "Why Are Modern Families Small? Toward an Evolutionary and Cultural Explanation for the Demographic Transition." *Personality and Social Psychology Review* 9:360–75.

Newson, Lesley, and Peter Richerson. 2009. "Why Do People Become Modern? A Darwinian Mechanism." *Population and Development Review* 35:117–58.

Nila, Sarah, Julien Barthes, Pierre-Andre Crochet, Bambang Suryobroto, and Michel Raymond. 2018. "Kin Selection and Male Homosexual Preference in Indonesia." *Archives of Sexual Behavior* 47 (8): 2455–65.

Nisbet, Ian C. T., and Jeremy J. Hatch. 1999. "Consequences of a Female-Biased Sex-Ratio in a Socially Monogamous Bird: Female-Female Pairs in the Roseate Tern Sterna Dougallii." *Ibis* 141 (2): 307–20.

Nisbett, Richard, and Sharon Gurwitz. 1970. "Weight, Sex, and the Eating Behavior of Human Newborns." *Journal of Comparative and Physiological Psychology* 73:245–53.

Ober, W. B. 1984. "The Sticky End of Frantisek Koczwara, Composer of 'The Battle of Prague.'" *American Journal of Forensic Medicine and Pathology* 5:142–50.

Odenwald, W. F., and S. D. Zhang. 1995. "Misexpression of the White (w) Gene Triggers Male-Male Courtship in Drosophila." *Proceedings of the National Academy of Sciences* 92 (12): 5525–29.

Olatunji, Bunmi. 2008. "Disgust, Scrupulosity and Conservative Attitudes about Sex: Evidence for a Mediational Model of Homophobia." *Journal of Research in Personality* 42:1364–69.

Oldham, Jeff, and Tim Kasser. 1999. "Attitude Change in Response to Information That Male Homosexuality Has a Biological Basis." *Journal of Sex and Marital Therapy* 25:121–24.

Olvera-Hernández, Sandra, Roberto Chavira, and Alonso Fernández-Guasti. 2015. "Prenatal Letrozole Produces a Subpopulation of Male Rats with Same-Sex Preference and Arousal as Well as Female Sexual Behavior." *Physiology & Behavior* 139 (February): 403–11.

Olyan, S. M. 1994. "'And with a Male You Shall Not Lie the Lying Down of a Woman': On the Meaning and Significance of Leviticus 18:22 and 20:13." *Journal of the History of Sexuality* 5:179–206.

Oosterhuis, Harry. 2000. *Stepchildren of Nature: Krafft-Ebing, Psychiatry, and the Making of Sexual Identity*. Chicago: University of Chicago Press.

Orrells, Daniel. 2015. *Sex: Antiquity and Its Legacy*. Oxford: Oxford University Press.

Ortiz, Daniel R. 1993. "Creating Controversy: Essentialism and Constructivism and the Politics of Gay Identity." *Virginia Law Review* 79:1833–57.

Osten-Sacken, Carl Robert von der. 1879. "Über einige Fälle von Copula inter Mares bei Insekten." *Entomologische Zeitung* 40:116–18.

Ovid. 2000. *Metamorphoses*. Translated by A. S. Kline. The Ovid Collection, University of Virginia Electronic Text Center. https://ovid.lib.virginia.edu/trans/Metamorph.htm.

Pappas, S. 2017. "Gay Lions? Not Quite." LiveScience, November 10. https://www.livescience.com/60910-gay-lions-not-quite.html.

Peakman, Julie. 2009. "Sexual Perversions in History: An Introduction." In *Sexual Perversions, 1670–1890*, edited by Julie Peakman, 1–49. New York: Palgrave MacMillan.

Peragallo, Alexandre. 1863. "Seconde note pour servir à l'histoire des lucioles." *Annales de la Société Entomologique de France* 4:661–65.

Perry, Mary. 1989. "The 'Nefarious Sin' in Early Modern Seville." In *The Pursuit of Sodomy: Male Homosexuality in Renaissance and Enlightenment Europe*, edited by Kent Gerard and Gert Hekma, 67–90. London: Haworth Press.

Peters, N. J. 2006. *Conundrum: The Evolution of Homosexuality*. Bloomington, IN: Authorhouse.

Peterson, David. 2012. "Where the Sidewalk Ends: The Limits of Social Constructionism." *Journal for the Theory of Social Behaviour* 42 (4): 465–84.

Pfaff, Donald. 1999. *Drive: Neural and Molecular Mechanisms for Sexual Motivation*. Cambridge, MA: MIT Press.

Pfaus, James G., Tod E. Kippin, and Genaro A. Coria-Avila. 2003. "What Can Animal Models Tell Us about Human Sexual Response?" *Annual Review of Sex Research* 14:1–63.

Pfaus, James G., Tod E. Kippin, Genaro A. Coria-Avila, Hélène Gelez, Veronica M. Afonso, Nafissa Ismail, and Mayte Parada. 2012. "Who, What, Where, When (and Maybe Even Why)? How the Experience of Sexual Reward Connects Sexual Desire, Preference, and Performance." *Archives of Sexual Behavior* 41 (1): 31–62.

Pfaus, James G., Mark F. Wilkins, Nina DiPietro, Michael Benibgui, Rachel Toledano, Anna Rowe, and Melissa Castro Couch. 2010. "Inhibitory and Disinhibitory Effects of Psychomotor Stimulants and Depressants on the Sexual Behavior of Male and Female Rats." *Hormones and Behavior* 58 (1): 163–76.

Pincemy, Gwénaëlle, F. Stephen Dobson, and Pierre Jouventin. 2010. "Homosexual Mating Displays in Penguins: Homosexual Mating Displays in Penguins." *Ethology* 116 (12): 1210–16.

Piskur, Julie, and Douglas Degelman. 1992. "Effects of Reading a Summary of Research about Biological Bases of Homosexual Orientation in Attitudes Toward Homosexuals." *Psychological Reports* 71:1219–25.

Plant, Richard. 2011. *The Pink Triangle: The Nazi War against Homosexuals*. New York: Macmillan.

Plato. 1925. *Symposium*. Translated by W. R. M. Lamb. Loeb Classical Library 166. Cambridge, MA: Harvard University Press.

Pliny the Elder. 1855. *The Natural History*. Translated by John Bostock. London: Taylor and Francis.

Plotnik, J. M., F. B. M. de Waal, and D. Reiss. 2006. "Self-Recognition in an Asian Elephant." *Proceedings of the National Academy of Sciences* 103 (45): 17053–57.

Plutarch. 1900. *Plutarch's Moralia*. Translated by Harold F. Cherniss and William C. Helmbold. Cambridge, MA: Harvard University Press.

Poiani, Aldo. 2010. *Animal Homosexuality: A Biosocial Perspective*. Cambridge: Cambridge University Press.

Polderman, Tinca J. C., Beben Benyamin, Christiaan A. de Leeuw, Patrick F. Sullivan, Arjen van Bochoven, Peter M. Visscher, and Danielle Posthuma. 2015. "Meta-Analysis of the Heritability of Human Traits Based on Fifty Years of Twin Studies." *Nature Genetics* 47 (7): 702–9.

Potts, Malcolm, and Roger V. Short. 1999. *Ever since Adam and Eve: The Evolution of Human Sexuality*. Cambridge: Cambridge University Press.

Pratarelli, Marc, and Jennifer Donaldson. 1997. "Immediate Effects of Written Material on Attitudes toward Homosexuality." *Psychological Reports* 81:1411–15.

Prior, Helmut, Ariane Schwarz, and Onur Güntürkün. 2008. "Mirror-Induced Behavior in the Magpie (Pica Pica): Evidence of Self-Recognition." Edited by Frans de Waal. *PLOS Biology* 6 (8): 1642–50.

Proctor, Robert. 1988. *Racial Hygiene: Medicine under the Nazis*. Cambridge, MA: Harvard University Press.

Rahman, Qazi, Anthony Collins, Martine Morrison, Jennifer Claire Orrells, Khatija Cadinouche, Sherene Greenfield, and Sabina Begum. 2008. "Maternal Inheritance and Familial Fecundity Factors in Male Homosexuality." *Archives of Sexual Behavior* 37 (6): 962–69.

Rahman, Qazi, and Glenn Wilson. 2003. "Born Gay? The Psychobiology of Human Sexual Orientation." *Personality and Individual Differences* 34:1337–82.

Ramagopalan, Sreeram V., David A. Dyment, Lahiru Handunnetthi, George P. Rice, and George C. Ebers. 2010. "A Genome-Wide Scan of Male Sexual Orientation." *Journal of Human Genetics* 55 (2): 131–32.

Ramirez, Luisa. 2007. "Dual Intergroup Meanings of Essentialism: Implications for Understanding Prejudice towards African Americans and Gay Men." PhD diss., Stony Brook University.

Ransom, Timothy W. 1981. *Beach Troop of the Gombe (The Primates)*. Lewisburg, PA: Bucknell University Press.

Rao, T. S. Sathyanarayana, and Chittaranjan Andrade. 2019. "Older Brothers and Male Homosexuality: Antibodies as an Explanation." *Journal of Psychosexual Health* 1 (2): 109–10.

Ratzsch, D., and J. Koperski. 2005. "Teleological Arguments for God's Existence." In *The Stanford Encyclopedia of Philosophy* (Summer 2019 edition), edited by E. Zalta. https://plato.stanford.edu/entries/teleological-arguments/.

Regan, Tom. 1986. *Bloomsbury's Prophet: G. E. Moore and the Development of His Moral Philosophy*. Eugene, OR: Wipf and Stock Publishers.

Regnerus, Mark. 2012. "How Different Are the Adult Children of Parents Who Have Same-Sex Relationships? Findings from the New Family Structures Study." *Social Science Research* 41:752–70.

Rellini, Alessandra H., Katie M. McCall, Patrick K. Randall, and Cindy M. Meston. 2005. "The Relationship between Women's Subjective and Physiological Sexual Arousal." *Psychophysiology* 42 (1): 116–24.

Reynolds, J. F. 1972. "On the Nesting Habits of White-Throated Bee-Eaters Merops Albicollis." *Bulletin of the East African Natural History Society*, 116–20.

Richerson, Peter, and Robert Boyd. 2005. *Not By Genes Alone: How Culture Transformed Human Evolution*. Chicago: University of Chicago Press.

Rieger, Gerulf, Ray Blanchard, Gene Schwartz, J. Michael Bailey, and Alan R. Sanders. 2012. "Further Data Concerning Blanchard's (2011) 'Fertility in the Mothers of Firstborn Homosexual and Heterosexual Men.'" *Archives of Sexual Behavior* 41 (3): 529–31.

Rimke, Heidi, and Alan Hunt. 2002. "From Sinners to Degenerates: The Medicalization of Morality in the 19th Century." *History of the Human Sciences* 15:59–88.

Roberts, Brian B. 1940. *The Breeding Behaviour of Penguins: With Special Reference to Pygoscelis Papua (Forster)*. Vol. 1. British Graham Land Expeditions 1934-37, Scientific Report. London: British Museum.

Rocke, Michael J. 1988. "Sodomites in Fifteenth-Century Tuscany: The Views of Bernardino of Siena." *Journal of Homosexuality* 16 (1–2): 7–31.

———. 1996. *Forbidden Friendships: Homosexuality and Male Culture in Renaissance Florence*. Oxford: Oxford University Press.

Rohy, Valerie. 2012. "On Homosexual Reproduction." *Differences* 23:101–30.

Rose, Steven Peter Russell, Richard Charles Lewontin, and Leon J. Kamin. 1984. *Not in Our Genes: Biology, Ideology and Human Nature*. New York: Pantheon Books.
Roselli, Charles E. 2020. "Programmed for Preference: The Biology of Same-Sex Attraction in Rams." *Neuroscience & Biobehavioral Reviews* 114 (July): 12–15.
Roselli, Charles E., John A. Resko, and Fred Stormshak. 2002. "Hormonal Influences on Sexual Partner Preference in Rams." *Archives of Sexual Behavior* 31:43–49.
Rotzoll, Maike, Wolfgang Eckart, Petra Fuchs, Geriff Hohendorf, and Christoph Mundt. 2010. *Die nationalsozialistische "Euthanasie"-Aktion "T-4" und ihre Opfer*. Paderborn, Germany: Ferdinand Schöningh.
Roughgarden, Joan. 2009. *Evolution's Rainbow: Diversity, Gender and Sexuality in Nature and People*. 2nd ed. Los Angeles: University of California Press.
———. 2017. "Homosexuality and Evolution." In *On Human Nature: Biology, Psychology, Ethics, Politics, and Religion*, edited by Michel Tibayrenc and Francisco J. Ayala, 495–516. Boston: Elsevier.
Rowlands, M. 2009. *Animal Rights: Moral Theory and Practice*. 2nd ed. Basingstoke, UK: Palgrave MacMillan.
Ruoso, Cyril. 2011. "Cold Embrace." Natural History Museum. https://www.nhm.ac.uk/wpy/gallery/2011-cold-embrace?tags=ed.10.
Rüsch, N., M. C. Angermeyer, and P. W. Corrigan. 2005. "Mental Illness Stigma: Concepts, Consequences, and Initiatives to Reduce Stigma." *European Psychiatry* 20:529–39.
Ruse, Michael. 2012. "Evolutionary Medicine." In *Evolution 2.0: Implications of Darwinism in Philosophy and the Social and Natural Sciences*, edited by Martin H. Brinkworth and Friedel Weinert, 177–89. Berlin: Springer.
Russell, Douglas G. D., William J. L. Sladen, and David G. Ainley. 2012. "Dr. George Murray Levick (1876–1956): Unpublished Notes on the Sexual Habits of the Adélie Penguin." *Polar Record* 48 (4): 387–93.
Ryle, Gilbert. 1949. *The Concept of Mind*. London: Hutchinson.
Ryne, Camilla. 2009. "Homosexual Interactions in Bed Bugs: Alarm Pheromones as Male Recognition Signals." *Animal Behaviour* 78 (6): 1471–75.
Saad, Lydia. 2018. "More Say 'Nature' Than 'Nurture' Explains Sexual Orientation." *Gallup*, May 24. https://news.gallup.com/poll/234941/say-nature-nurture-explains-sexual-orientation.aspx.
Sakalli, Nuray. 2002. "Application of the Attribution-Value Model of Prejudice to Homosexuality." *Journal of Social Psychology* 142:264–71.
Samuels, Richard. 2002. "Nativism in Cognitive Science." *Mind & Language* 17:233–65.
Sanders, Alan R., Gary W. Beecham, Shengru Guo, Khytam Dawood, Gerulf Rieger, Judith A. Badner, Elliot S. Gershon, et al. 2017. "Genome-Wide Association Study of Male Sexual Orientation." *Scientific Reports* 7 (1): 16950.
Sanders, Alan R., Eden R. Martin, Gary W. Beecham, Shengru Guo, K. Dawood, G. Rieger, J. A. Badner, et al. 2015. "Genome-Wide Scan Demonstrates Significant Linkage for Male Sexual Orientation." *Psychological Medicine* 45 (7): 1379–88.
Sandfort, Theo G. M. 2005. "Sexual Orientation and Gender: Stereotypes and Beyond." *Archives of Sexual Behavior* 34 (6): 595–611.
Savoia, Paulo. 2010. "Sexual Science and Self-Narrative: Epistemology and Narrative Technologies of the Self between Krafft-Ebing and Freud." *History of the Human Sciences* 23:1–25.

Savolainen, Vincent, and Jason A. Hodgson. 2016. "Evolution of Homosexuality." In *Encyclopedia of Evolutionary Psychological Science*, edited by Viviana Weekes-Shackelford and Todd K. Shackelford, 1–8. Cham, Switzerland: Springer.

Schopenhauer, Arthur. 2018. *The World as Will and Representation*. Edited by Christopher Janaway. Translated by Judith Norman and Alistair Welchman. Cambridge: Cambridge University Press.

Schüklenk, Udo, Edward Stein, Jacinta Kerin, and William Byne. 1997. "The Ethics of Genetic Research on Sexual Orientation." *Hastings Center Report* 27 (4): 6–13.

Schulz, Jonathan F., Duman Bahrami-Rad, Jonathan P. Beauchamp, and Joseph Henrich. 2019. "The Church, Intensive Kinship, and Global Psychological Variation." *Science* 366 (6466): eaau5141.

Schwartz, Gene, Rachael M. Kim, Alana B. Kolundzija, Gerulf Rieger, and Alan R. Sanders. 2010. "Biodemographic and Physical Correlates of Sexual Orientation in Men." *Archives of Sexual Behavior* 39 (1): 93–109.

Schwitzgebel, Eric. Forthcoming. "The Pragmatic Metaphysics of Belief." In *The Fragmented Mind*, edited by Cristina Borgoni, Dirk Kindermann, and Andrea Onofri. Oxford: Oxford University Press.

Scott, Isabel M., Andrew P. Clark, Steven C. Josephson, Adam H. Boyette, Innes C. Cuthill, Ruby L. Fried, Mhairi A. Gibson, et al. 2014. "Human Preferences for Sexually Dimorphic Faces May Be Evolutionarily Novel." *Proceedings of the National Academy of Sciences* 111 (40): 14388–93.

Scott, Sydney E., and Paul Rozin. 2020. "Actually, Natural Is Neutral." *Nature Human Behaviour* 4:989–90.

Sedgwick, Eve Kosofsky. 1990. *Epistemology of the Closet*. Berkeley: University of California Press.

———. 2008. *Epistemology of the Closet*. Updated with a new preface. Berkeley: University of California Press.

Seitler, Dana. 2004. "Queer Physiognomies: Or How Many Ways Can We Do the History of Sexuality?" *Criticism* 46:71–102.

Sesardic, Neven. 2005. *Making Sense of Heritability*. Cambridge: Cambridge University Press.

Shanks, Niall, Ray Greek, and Jean Greek. 2009. "Are Animal Models Predictive for Humans?" *Philosophy, Ethics, and Humanities in Medicine* 4 (1): 2.

Shorter, Edward. 1997. *A History of Psychiatry from the Era of the Asylum to the Era of Prozac*. New York: John Wiley & Sons.

———. 2008. *Before Prozac: The Troubled History of Mood Disorders in Psychiatry*. New York: Oxford University Press.

Shortland, Michael. 1987. "Courting the Cerebellum: Early Organological and Phrenological Views of Sexuality." *British Journal for the History of Science* 20:173–99.

Silverman, Hugh. 2000. *Philosophy and Desire*. New York: Routledge.

Singer, Barry. 1984. "Conceptualizing Sexual Arousal and Attraction." *Journal of Sex Research* 20:230–40.

Skirbekk, Vegard. 2008. "Fertility Trends by Social Status." *Demographic Research* 18 (March): 145–80.

Slack, Nancy. 2010. *Evelyn Hutchinson and the Invention of Modern Ecology*. New Haven, CT: Yale University Press.

Smith, Brianna A., Zein Murib, Matthew Motta, Timothy H. Callaghan, and Marissa Theys. 2018. "'Gay' or 'Homosexual'? The Implications of Social Category Labels for the Structure of Mass Attitudes." *American Politics Research* 46 (2): 336–72.

Smith, D. L. 2014. "Dehumanization, Essentialism, and Moral Psychology." *Philosophy Compass* 9:814–24.

Smith, Dinitia. 2004. "Love That Dare Not Squeak Its Name." *New York Times*, February 7. https://www.nytimes.com/2004/02/07/arts/love-that-dare-not-squeak-its-name.html.

Smith, Glenn, Annie Bartlett, and Michael King. 2004. "Treatments of Homosexuality in Britain since the 1950s—An Oral History: The Experience of Patients." *British Medical Journal* 328:427–29.

Smith, Michael. 1994. *The Moral Problem*. Hoboken, NJ: Wiley-Blackwell.

Smuts, Barbara. 1985. *Sex and Friendship in Baboons*. Piscataway, NJ: Transaction Publishers.

Smuts, Barbara, and John Watanabe. 1990. "Social Relationships and Ritualized Greetings in Adult Male Baboons (Papio Cynocephalus Anubis)." *International Journal of Primatology* 11:147–72.

Soble, Alan. 2008. *The Philosophy of Sex and Love*. 2nd ed. St. Paul, MN: Paragon House.

Søli, Geir. 2009. "Against Nature? An Exhibition on Animal Homosexuality." Naturhistorisk Museum. http://www.nhm.uio.no/besok-oss/utstillinger/skiftende/againstnature/index-eng.html.

Solomon, Robert. 1975. "Sex and Perversion." In *Philosophy and Sex*, edited by R. Baker and F. Elliston. Buffalo, NY: Prometheus Books.

Sommer, Volker, Peter Schauer, and Diana Kyriazis. 2006. "A Wild Mixture of Motivations: Same-Sex Mounting in Indian Langur Monkeys." In *Homosexual Behaviour in Animals: An Evolutionary Perspective*, edited by Volker Sommer and Paul L. Vasey, 238–72. Cambridge: Cambridge University Press.

Sommer, Volker, and Paul L. Vasey. 2006. *Homosexual Behavior in Animals: An Evolutionary Perspective*. Cambridge: Cambridge University Press.

Spitzer, R. L., and J. C. Wakefield. 1999. "DSM-IV Diagnostic Criterion for Clinical Significance: Does It Help Solve the False Positives Problem?" *American Journal of Psychiatry* 156 (12): 1856–64.

Spitzer, Robert. 1973. "A Proposal about Homosexuality and the APA Nomenclature: Homosexuality as an Irregular Form of Sexual Behavior and Sexual Orientation Disturbance as a Psychiatric Disorder." *American Journal of Psychiatry* 130:1214–16.

———. 1981. "The Diagnostic Status of Homosexuality in DSM-III: A Reformulation of the Issues." *American Journal of Psychiatry* 138:210–15.

———. 2001. "Values and Assumptions in the Development of DSM-III and DSM-III-R: An Insider's Perspective and a Belated Response to Sadler, Hulgus, En Agich's 'On Values in Recent American Psychiatric Classification.'" *Journal of Nervous and Mental Disease* 189:351–59.

———. 2003. "Can Some Gay Men and Lesbians Change Their Sexual Orientation? 200 Participants Reporting a Change from Homosexual to Heterosexual Orientation." *Archives of Sexual Behavior* 32:403–17.

Stegenga, Jacob. 2018. *Care and Cure: An Introduction to Philosophy of Medicine*. Chicago: University of Chicago Press.

Stein, Edward. 1999. *The Mismeasure of Desire: The Science, Theory, and Ethics of Sexual Orientation*. Oxford: Oxford University Press.

Stengers, Isabelle. 2000. *The Invention of Modern Science*. Translated by Daniel W. Smith. Minneapolis: University of Minnesota Press.

Stephens, Elizabeth. 2008. "Pathologizing Leaky Male Bodies: Spermatorrhea in Nineteenth-Century British Medicine and Popular Anatomical Museums." *Journal of the History of Sexuality* 17:421–38.

Stoller, Robert, Irving Bieber, Ronald Gold, Richard Green, Judd Marmor, Charles Socarides, and Robert Spitzer. 1973. "A Symposium: Should Homosexuality Be in the APA Nomenclature?" *American Journal of Psychiatry* 130:1207–16.

Strawson, Galen. 1994. *Mental Reality*. Cambridge, MA: MIT Press.

Strawson, Peter F. 1985. *Skepticism and Naturalism: Some Varieties*. London: Methuen.

Sulloway, Frank. 1979. *Freud, Biologist of the Mind: Beyond the Psychoanalytic Legend*. Cambridge, MA: Harvard University Press.

Swaab, D. F. 2004. "Sexual Differentiation of the Human Brain: Relevance for Gender Identity, Transsexualism and Sexual Orientation." *Gynecological Endocrinology* 19 (6): 301–12.

———. 2007. "Sexual Differentiation of the Brain and Behavior." *Best Practice & Research Clinical Endocrinology & Metabolism* 21 (3): 431–44.

Swift-Gallant, Ashlyn. 2019. "Individual Differences in the Biological Basis of Androphilia in Mice and Men." *Hormones and Behavior* 111 (May): 23–30.

Takács, Judit. 2004. "The Double Life of Kertbeny." In *Past and Present of Radical Sexual Politics*, edited by Gert Hekma, 51–62. Amsterdam: UvA–Mosse Foundation.

Talhelm, Thomas. 2018. "Hong Kong Liberals Are WEIRD: Analytic Thought Increases Support for Liberal Policies." *Personality and Social Psychology Bulletin* 44 (5): 717–28.

Talisse, R. B., and S. F. Aikin. 2007. "Kitcher on the Ethics of Inquiry." *Journal of Social Philosophy* 38:654–65.

Tennent, W. John. 1987. "A Note on the Apparent Lowering of Moral Standards in the Lepidoptera." *Entomologist's Record and Journal of Variation* 99:81–83.

Teodorov, Elizabeth, Simone Angelica Salzgeber, Luciano F. Felicio, Franci M. F. Varolli, and Maria Martha Bernardi. 2002. "Effects of Perinatal Picrotoxin and Sexual Experience on Heterosexual and Homosexual Behavior in Male Rats." *Neurotoxicology and Teratology* 24 (2): 235–45.

Terry, Jennifer. 2000. "'Unnatural Acts' in Nature: The Scientific Fascination with Queer Animals." *GLQ: A Journal of Lesbian and Gay Studies* 6:151–93.

Thorp, John. 1992. "The Social Construction of Homosexuality." *Phoenix* 46:54–61.

Tovee, M., V. Swami, A. Furnham, and R. Mangalparsad. 2006. "Changing Perceptions of Attractiveness as Observers Are Exposed to a Different Culture." *Evolution and Human Behavior* 27 (6): 443–56.

Triana-Del Rio, Rodrigo, Miriam B. Tecamachaltzi-Silvarán, Victor X. Díaz-Estrada, Deissy Herrera-Covarrubias, Aleph A. Corona-Morales, James G. Pfaus, and Genaro A. Coria-Avila. 2015. "Conditioned Same-Sex Partner Preference in Male Rats Is Facilitated by Oxytocin and Dopamine: Effect on Sexually Dimorphic Brain Nuclei." *Behavioural Brain Research* 283 (April): 69–77.

Trivers, Robert L. 1985. *Social Evolution*. Menlo Park, CA: Benjamin/Cummings.

Troiden, Richard. 1985. "Self, Self-Concept, Identity, and Homosexual Identity: Constructs in Need of Definition and Differentiation." *Journal of Homosexuality* 10:97–109.

———. 1988. "Homosexual Identity Development." *Journal of Adolescent Healthcare* 9:105–13.

Trubody, Ben. 2016. "Richard Feynman's Philosophy of Science." *Philosophy Now* 114:10–12.

Trumbach, Randolph. 1998. *Sex and the Gender Revolution*. Vol. 1, *Heterosexuality and the Third Gender in Enlightenment London*. Chicago: University of Chicago Press.

Tulchin, Allan. 2007. "Same-Sex Couples Creating Household in Old Regime France: The Uses of the Affrèrement," *Journal of Modern History* 79 (3): 613–47.

Turkheimer, Eric. 2000. "Three Laws of Behavior Genetics and What They Mean." *Current Directions in Psychological Science* 9:160–64.

Ulrichs, Karl Heinrich. 1994. *The Riddle of "Man-Manly" Love: The Pioneering Work on Male Homosexuality in Two Volumes*. Translated by Michael Lombardi-Nash. New York: Prometheus Books.

Utzeri, Carlo, and Carlo Belfiore. 1990. "Tandem anomali fra Odonati (Odonata)." *Fragmenta Entomologica* 22:271–87.

VanderLaan, Doug P., Zhiyuan Ren, and Paul L. Vasey. 2013. "Male Androphilia in the Ancestral Environment: An Ethnological Analysis." *Human Nature* 24 (4): 375–401.

VanderLaan, Doug P., and Paul L. Vasey. 2011. "Male Sexual Orientation in Independent Samoa: Evidence for Fraternal Birth Order and Maternal Fecundity Effects." *Archives of Sexual Behavior* 40 (3): 495–503.

———. 2014. "Evidence of Cognitive Biases for Maximizing Indirect Fitness in Samoan Fa'afafine." *Archives of Sexual Behavior* 43:1009–22.

Van der Meer, Theo. 1989. "The Persecutions of Sodomites in Eighteenth-Century Amsterdam: Changing Perceptions of Sodomy." In *The Pursuit of Sodomy: Male Homosexuality in Renaissance and Enlightenment Europe*, edited by Kent Gerard and Gert Hekma, 263–310. London: Haworth Press.

Van de Ven, Paul, Pamela Rodden, June Crawford, and Susan Kippax. 1997. "A Comparative Demographic and Sexual Profile of Older Homosexually Active Men." *Journal of Sex Research* 34 (4): 349–60.

Van Gossum, Hans, Luc De Bruyn, and Robby Stoks. 2005. "Reversible Switches between Male-Male and Male-Female Mating Behaviour by Male Damselflies." *Biology Letters* 1:268–70.

Van Gossum, Hans, T. Robb, M. R. Forbes, and L. Rasmussen. 2008. "Female-Limited Polymorphism in a Widespread Damselfly: Morph Frequencies, Male Density, and Phenotypic Similarity of Andromorphs to Males." *Canadian Journal of Zoology* 86 (10): 1131–38.

Vasey, Paul L. 1995. "Homosexual Behavior in Primates: A Review of Evidence and Theory." *International Journal of Primatology* 16:173–204.

———. 2002. "Same-Sex Sexual Partner Preference in Hormonally and Neurologically Unmanipulated Animals." *Annual Review of Sex Research* 13:141–79.

Vasey, Paul L., Jessica L. Parker, and Doug P. VanderLaan. 2014. "Comparative Reproductive Output of Androphilic and Gynephilic Males in Samoa." *Archives of Sexual Behavior* 43 (2): 363–67.

Vasey, Paul L., and Volker Sommer. 2006. "Homosexual Behaviour in Animals: Topics, Hypotheses and Research Trajectories." In *Homosexual Behaviour in Animals: An Evolutionary Perspective*, edited by Volker Sommer and Paul Vasey, 3–44. Cambridge: Cambridge University Press.

Vasey, Paul L., and Doug P. VanderLaan. 2010. "Avuncular Tendencies and the Evolution of Male Androphilia in Samoan Fa'afafine." *Archives of Sexual Behavior* 39 (4): 821–30.

———. 2012. "Sexual Orientation in Men and Avuncularity in Japan: Implications for the Kin Selection Hypothesis." *Archives of Sexual Behavior* 41 (1): 209–15.

Wakefield, Jerome C. 1992. "The Concept of Mental Disorder: On the Boundary between Biological Facts and Social Values." *American Psychologist* 47:232–47.

———. 1999. "Harmful Dysfunction and the DSM Definition of Mental Disorder." *Journal of Abnormal Psychology* 108:430–32.

———. 2013. "Addiction, the Concept of Disorder, and Pathways to Harm: Comment on Levy." *Frontiers in Psychiatry* 4:34.

Wakefield, Jerome C., and Jordan A Conrad. 2020. "Harm as a Necessary Component of the Concept of Medical Disorder: Reply to Muckler and Taylor." *Journal of Medicine and Philosophy* 45 (3): 350–70.

Walker, Robert, Michael Gurven, Kim Hill, Andrea Migliano, Napoleon Chagnon, Roberta De Souza, Gradimir Djurovic, et al. 2006. "Growth Rates and Life Histories in Twenty-Two Small-Scale Societies." *American Journal of Human Biology* 18 (3): 295–311.

Wallis, Faith. 2008. "Giulio Guastavini's Commentary on Pseudo-Aristotle's Account of Male Same-Sexual Coitus, Problemata 4.26." In *The Sciences of Homosexuality in Early Modern Europe*, edited by Kenneth Borris and George Rousseau, 57–74. London: Routledge.

Wang, Y., and M. Kosinski. 2018. "Deep Neural Networks Are More Accurate Than Humans at Detecting Sexual Orientation from Facial Images." *Journal of Personality and Social Psychology* 114:246–57.

Waters, Chris. 2006. "Sexology." In *Palgrave Advances in the Modern History of Sexuality*, edited by Harry Cocks and Matt Houlbrook, 41–63. London: Palgrave MacMillan.

Watkins, Susan. 1990. "From Local to National Communities: The Transformation of Demographic Regimes in Western Europe, 1870–1960." *Population and Development Review* 16:241–72.

Weber, Matthias. 2000. "Psychiatric Research and Science Policy in Germany. The History of the Deutsche Forschungsanstalt Für Psychiatrie (German Institute for Psychiatric Research) in Munich from 1917 to 1945." *History of Psychiatry* 11:235–58.

Weeks, Jeffrey. 2000. *Making Sexual History*. Oxford: Polity Press.

———. 2007. *The World We Have Won: The Remaking of Erotic and Intimate Life*. Abingdon, UK: Routledge.

Weindling, Paul. 1989. *Health, Race and German Politics between National Unification and Nazism, 1870–1945*. Cambridge: Cambridge University Press.

Whitham, Jessica, and Dario Maestripieri. 2003. "Primate Rituals: The Function of Greetings between Male Guinea Baboons." *Ethology* 109:847–59.

Wickler, Wolfgang. 1967. "Socio-Sexual Signals and Their Intra-Specific Initiation among Primates." In *Primate Ethology*, edited by Desmond Morris, 89–189. Chicago: Aldine.

Williams, Craig A. 2010. *Roman Homosexuality*. Oxford: Oxford University Press.

Williams, Melissa, and Jennifer Ebenhardt. 2008. "Biological Conceptions of Race and the Motivation to Cross Racial Boundaries." *Journal of Personality and Social Psychology* 94:1033–47.

Wilson, Edward. 1975. *Sociobiology: The New Synthesis*. Cambridge, MA: Harvard University Press.

———. 1978. *On Human Nature*. Cambridge, MA: Harvard University Press.

Yankelovich Partners. 1994. "A Yankelovich MONITOR Perspective on Gays/Lesbians." Norwalk, CT: Yankelovich Partners.

Young, Lindsay, Eric VanderWerf, and Brenda Zaun. 2008. "Successful Same-Sex Pairing in Laysan Albatross." *Biology Letters* 4:323–25.

Yzerbyt, Vincent, Stephan Rocher, and Georges Schadron. 1997. "Stereotypes as Explanations: A Subjective Essentialistic View of Group Perception." In *The Psychology of Stereotyping and Group Life*, edited by Naomi Ellemers, Alexander Haslam, Russell Spears, and Penelope Oakes, 20–50. Oxford: Basil Blackwell.

Zahavi, Amotz. 1977. "Reliability in Communication Systems and the Evolution of Altruism." In *Evolutionary Ecology*, edited by Bernard Stonehouse and Christopher Perrins, 253–59. London: Palgrave MacMillan.

Zietsch, B., K. Morley, S. Shekar, K. Verweij, M. Keller, S. Macgregor, M. Wright, J. Bailey, and N. Martin. 2008. "Genetic Factors Predisposing to Homosexuality May Increase Mating Success in Heterosexuals." *Evolution and Human Behavior* 29 (6): 424–33.

Zuk, Marlene. 2006. "Family Values in Black and White." *Nature* 439:917.

———. 2011. *Sex on Six Legs: Lessons on Life, Love and Language from the Insect World*. New York: Houghton Mifflin Harcourt.

Index

adaptation, 33, 42, 123, 128, 130, 133
adultery, 64, 177
Afghanistan, 117
Ainley, David, 64
alcoholism, 24, 27, 30f, 142
alliance formation hypothesis, 81, 100, 123–28, 201n8
Allison, Anthony, 105
ambisexual, 90
American Psychiatric Association (APA), 151–61, 164
anal penetration, 55–56, 118, 141, 147, 162, 176, 198n4, 199n18
androgens, 37, 94, 95, 97, 98, 185
animal homosexual preference, 7, 67, 72, 74, 94
anthropology, 2, 5, 8, 12, 49, 51, 87, 89, 100, 114–15, 117, 122, 123, 125, 127, 129, 130, 135, 149, 195n5, 200n19, 201n2
anthropomorphism, 51, 68, 72, 76–77, 79, 88, 89, 91, 200n22
antiquity, 1, 12, 17–18, 20, 23, 43–45, 55, 61, 115–18, 122, 124, 142, 176
APA (American Psychiatric Association), 151–61, 164
Aquinas, Thomas, 177
Aristotelian tradition, 1, 18, 19, 24, 114, 117, 200n25
Aristotle, 1, 61
arousal, 40, 69, 71, 82–83, 85–86, 146, 150, 158–59, 198n9, 199n18
auditory stimuli, 185
aversion, 18, 40, 176
Avicenna, 19

bacha bazi, 117, 118
Bagemihl, Bruce, 59–60, 63, 75, 79, 81, 85, 86, 89–90, 91, 93, 196n9, 198n7, 199n14, 199n17, 200n20, 200n21, 200n23
Beach, Frank, 94, 102, 200n25, 202n6
beak-genital propulsion, 81
bedbugs, 64
behavioral genetics, 12, 34, 35, 41
Bering, Jesse, 120
bestiality, 137, 146, 147. *See also* zoophilia
bird, 4, 17, 50, 61, 65, 69, 70, 71, 75, 109, 198n11
bisexual, 33, 90, 122, 123, 154, 195n7, 202n6
Blanchard, Ray, 38, 39, 113–15, 197n10
Boorse, Christopher, 9, 154, 163, 165, 167, 168, 169, 203n13
Boyd, Robert, 129
boy-inseminating practices, 8, 87–88
Bryant, Anita, 14
Buffon, Georges-Louis Leclerc de, 51, 140
Burley, Walter, 16, 19

canalization, 33, 36–41, 49, 197n9
castration, 25, 31, 148, 196n2
causality, 28, 36, 38, 45, 96, 181, 184, 197n12
celibacy, 109, 154, 155
Chang, Hasok, 3
Charcot, Jean-Martin, 142
Chauncey, George, 45
Christian: authors, 11; church, 133; commentators, 18; emperors, 18; fathers, 61; philosophers, 19; sexual prescriptions, 134; theologians, 176

Christianity, 18, 176, 177
cinaedus. See *kinaidoi*
cockchafers, 13, 51, 52f, 52–57, 60, 63, 64, 68, 70–73, 78, 81, 198n2. See also *Melolontha*
coitus, 52, 146, 147
concentration camps, 25, 31
conceptual analysis, 4, 32, 86, 167–73
concordance, 26, 27, 29, 34–36, 38, 196n3
conservativism, 89, 160, 178
constructivism, 12, 42–48, 49, 116, 127, 128, 133, 197nn12–14, 202n1. *See also* social construction
conversion therapy, 37, 40, 149, 153, 203n11
copulation, 13, 51–57, 64, 73, 75f, 79, 81, 92, 145, 199n17, 200n21
cross-cultural universals, 22, 23, 44, 45, 96
cystic fibrosis, 33, 34, 42

Davidson, Arnold, 145
degeneration, 23, 59, 63, 137, 138, 142–43, 198n6
Denmark, 162
desire (homosexual), 42–44, 46, 48, 50, 51, 54, 55, 57, 58, 60, 61, 62, 67, 68, 69–72, 73, 75, 76, 77, 78, 79, 81, 83, 87, 88, 93, 94, 118, 123, 127, 134, 138, 141, 142, 145, 146, 148, 150, 154, 158, 163, 169, 176, 195n5, 197n13, 198n12, 199n13, 202n1
Deutsche Forschungsanstalt für Psychiatrie (German Institute for Psychiatric Research), 24, 25
Diagnostic and Statistical Manual of Mental Disorders (DSM), 13, 151, 155, 156, 160–62, 202n9; *DSM-I*, 152–53; *DSM-II*, 152–53, 155, 157–59; *DSM-III*, 153, 155–56, 157–59; *DSM-III-R*, 159, 161; *DSM-IV*, 160; *DSM-IV-TR*, 160
diddling, 82–87, 126, 200n19
diet, 19, 34, 39
disease criteria: action, 159–61; distress, 152–55, 157, 158, 159, 160, 161; *DSM* criterion A, 158, 159; *DSM* criterion B, 159, 160; exclusivity, 147, 158–59, 160, 202n3; harm, 159, 161, 163, 164, 165, 167, 168, 171, 172, 177; impairment, 155, 157, 159, 160, 161
disgust, 59, 130, 196n8
disorder view of homosexuality, 3, 27, 40, 139, 154, 161–67
DNA, 27, 31, 104, 105, 109, 113
dominance, 53, 77–79, 80
DSM. See *Diagnostic and Statistical Manual of Mental Disorders (DSM)*
dysfunction, 93, 140, 157, 161, 163–65, 168, 202n9

ecology, 94, 98, 99, 101, 102
economic modernization, 131–34
"Ego-Dystonic Homosexuality," 159
eliminativism, 170–73
Ellis, Henry Havelock, 7, 63, 149, 150, 154, 202n4
endocrinology, 2, 12, 28, 36, 49, 65, 67, 92, 94, 96, 100, 114, 125, 181, 185
entomology, 13, 14, 51, 53, 54, 57, 63, 78, 92, 108, 137, 175
environment, 16, 26, 31, 34–38, 39, 41, 47, 107, 113, 128, 134, 177, 188, 196nn6–8
erastai, 44, 115, 116f, 122
Ereshefsky, Marc, 171
eromenoi, 44, 115, 116f, 122
error hypothesis. *See under* explanatory hypotheses (of animal homosexuality)
essentialism, 47, 48, 116, 118, 178, 179, 180, 181, 197n15, 204n3
essentialist beliefs, 178, 179–81
evolutionary biology, 12, 100, 101, 106, 111, 119, 123, 127, 128, 131, 135
evolutionary paradox (of homosexuality), 13, 99, 100, 101, 104, 123, 183
explanatory hypotheses (of animal homosexuality): error, 53–54, 60, 63–68, 72, 74, 96, 200n24; Hobson's choice, 60–63, 65, 66, 67, 72, 73, 74

fa'afaine, 110, 111, 113, 115, 121, 122
Facebook, 186–87
FBOE. *See* fraternal birth order effect (FBOE)
female homosexuality, 11, 12, 102, 201n6
fetishism, 3, 137, 146, 147, 153, 156, 157, 158
fetus, 38, 39, 41, 94, 191

fibromyalgia, 165
fireflies, 51, 56
fixation, 147, 148
fixity, 33, 34, 36, 37, 39, 40, 42, 180
folk psychology, 71
Foucault, Michel, 22, 45, 46, 145
France, 22, 51, 52, 53, 55, 66, 118, 137, 138, 139, 141, 150, 198n3, 201n5
fraternal birth order effect (FBOE), 11, 38, 39, 112, 113, 114, 197n10
Freud, Sigmund, 31, 33, 48, 92, 146–50, 153, 158, 202n3

Gall, Franz Joseph, 141
gay: activists, 6, 9, 20, 57, 58, 150, 154, 155, 166, 182, 188, 198n7, 202n4; as a characteristic of mutant fruit flies, 70, 77, 93, 94, 199n16, 200n22; as a human sexual identity category, 30, 36, 38, 74–77, 89, 95, 114, 134, 180, 188, 199n16, 204n5; liberation, 14, 101; as a replacement for "homosexual," 9–10; rights, 20; sciences, 2, 4, 6, 12, 14, 95, 119, 137, 175, 176, 182, 195n2; studies, 195n2
gaydar, 175, 185–89, 204n6
Gay Liberation Movement, 101
Gelman, Susan, 179
gene, 16, 26, 30, 31, 32, 33, 106, 108, 113, 114, 129, 180, 183, 196n7, 200n22
genetics: behavioral, 41, 130; of homosexuality, 5, 24, 25, 27, 35, 41, 49, 100, 101, 104, 119, 175, 195n5; of schizophrenia, 25
genome-wide association studies (GWAS), 31, 32, 34, 38, 113, 196n5
genotype, 38, 112
Germany, 6, 13, 14, 20, 24, 25, 31, 53, 55, 150, 175, 182, 203n16
Greece, 1, 11, 17, 18, 44, 45, 98, 115, 117, 118, 125, 176, 177
Griffiths, Paul, 42, 177
GWAS. *See* genome-wide association studies (GWAS)

Halperin, David, 43, 44, 46, 47, 116, 117, 183, 197n14
harm principle, 136

Haslam, Nick, 180
Henrich, Joseph, 133, 134, 182
heritability, 33–36, 41, 49, 106, 108, 182
heterosexuality, 2, 8, 22, 45, 47, 91, 97, 108, 135, 147, 154, 155, 159, 167, 183
heterozygotes, 104, 105, 106, 107
Hirschfeld, Magnus, 20, 21f, 22, 23, 24, 31, 33, 41, 149, 150, 154, 181, 182, 202n4
Hobson's choice hypothesis. *See under* explanatory hypotheses (of animal homosexuality)
homonegativity, 14, 15, 23, 57, 59, 136, 137, 154, 164, 167, 175, 176, 178, 180, 181, 182, 183, 184, 185, 186, 188, 191, 196n8, 204n5
homophobia, 100, 183, 196n8, 204n5
homosexual: act, 1, 18, 19, 23, 53, 54, 55, 56, 64, 141, 145, 146, 150, 162, 169, 202n7; copulation, 13, 51, 52, 53, 54, 55, 56, 57, 64, 73, 75f, 79, 81, 92, 199n17, 200n21. *See also* animal homosexual preference; desire (homosexual); identity (homosexual); preference (homosexual); sexual orientation
homosexuality: causes of, 1, 47, 48, 49, 51, 65, 97, 112, 114, 123, 142, 143, 145, 148, 150, 154, 175, 181, 184, 191, 195n5, 196n8, 197n15, 197n16, 204n10; congenital, 56, 58, 59, 198n6; development of, 27, 28, 29, 33, 38, 39, 40, 45, 47, 48, 94, 96, 97, 99, 103, 113, 114, 119, 148, 155, 170, 196n7; dimensions of, 13, 50, 60, 96, 112, 118, 123, 134, 195n5; as disability, 136; "Ego-Dystonic," 159; as mental disorder (*see* disorder view of homosexuality); modern, 8, 43, 46, 47, 88, 91, 100, 108, 112–23, 127, 128, 131, 132, 133, 134, 135, 145, 195n5; passive, 29, 44, 55, 64, 95–96, 114–15, 200n24; transgenerational or ritualized, 8, 18, 55, 115
homozygotes, 104–6, 108
hormones. *See* endocrinology
Hull, David, 135, 183
Hume, David, 177
Huntington's disease, 33
Huxley, Julian, 106

hybridism, 163–65
hypermasculinity, 94

identity (homosexual), 117, 118, 131, 133–35, 145, 174, 199n16
immune system, 38, 113–14
impermissible research, criteria for, 184
individualism, 133–34, 182
innateness, 4, 6, 7, 12, 13, 16–23, 31–42, 48, 49, 51, 56–58, 147, 149, 180, 197n11, 198n8
intersex conditions, 37
intrasexual, 90
intrauterine hormonal exposure, 38–39, 113, 185
isosexual, 8, 10, 87–90
Italy, 1, 63, 107, 110, 117

Kallmann, Franz, 14, 24–31, 34, 40, 101, 103
Kertbeny, Károly, 6–10, 20
kinaidoi, 117, 118
kin selection theory, 108–9, 111
Kinsey, Alfred, 102, 103, 150, 152
Kitcher, Philip, 182, 184, 189, 190, 204n4
Kourany, Janet, 182, 190, 204n7
Kraepelin, Emil, 24, 31, 150
Krafft-Ebing, Richard von, 14, 142–47, 150, 159, 161, 166, 202n2, 202n7

Lady Gaga, 9, 16
langur monkeys, 124
Laqueur, Thomas, 11
LeVay, Simon, 39, 75, 92
Levick, George, 64
LGBTQ, 9, 136, 186, 188
Linquist, Stefan, 42
Lombroso, Cesare, 142
lordosis, 94–97
low mood, 172

Machery, Edouard, 33, 42
Magnan, Valentin, 142, 143
Mameli, Matteo, 32, 36
Mamluks, 117, 124
marriage, 101, 122, 125, 130, 133, 134, 177, 180, 198n6, 201n5, 202nn4–5

masochism, 146–48, 153, 158
maybugs. *See* cockchafers; *Melolontha*
Mayr, Ernst, 106
medicalization of homosexuality, 146, 147, 172
medieval. *See* Middle Ages
Melolontha, 51; *hippocastani*, 53; *vulgaris*, 53
Melolontha story, 57, 58, 60, 63, 64, 67, 78, 81, 89, 91, 92, 198n7
mental disorder, 2, 13, 14, 24, 26, 58, 102, 136, 137, 138, 143, 151, 152, 153, 155, 157, 158, 159, 160, 161, 164, 165, 167, 196n8
Michael, George, 36
Middle Ages, 1, 2, 12, 18, 19, 20, 23, 117, 118, 124, 133, 134, 175
"mollies," 115
moral realism, 165, 203n10
Muccioli, Alessandro, 63

narcissism, 148, 158, 184
nativism, 12, 16, 17, 42–49. *See also* innateness
naturalist view of disorder, 161–65, 167, 168, 172–73
natural law, 176–78
naturalness of homosexuality, 12, 13, 17, 51, 58, 98, 198n7, 200n23
natural selection, 13, 98, 99, 101, 104, 106–8, 111, 123, 128, 129, 162, 183
Nazism, 24, 31, 182
Netherlands, 162
neuroscience, 12, 103
New Testament, 176
normativist view of disorder, 161–64, 169, 172, 203nn12–13

Odysseus, 61
Oedipus, 146, 148
Old Testament, 137, 176
orientation. *See* sexual orientation
Osmond, Humphry, 106
Osten-Sacken, Carl Robert, 54–57, 199n17

Papua New Guinea, 87, 115, 124, 125, 200n19
paraphilia, 157–60. *See also* sexual perversion

pathologization of homosexuality, 136, 137, 165–66. *See also* medicalization of homosexuality
pederasty, 6, 23, 55–56, 61, 63, 67, 72, 115, 117, 122, 198n4
pedophilia, 3, 137, 146, 147, 153, 157
penguins, 64–65, 89
Peragallo, Alexandre, 51
Per scientiam ad justitiam, 20, 21f, 149, 181
phenotype, 8, 34, 38, 100, 112, 114, 119, 163, 195n6, 196n8, 197n15
phrenology, 141
polymorphism, 103–6, 108
population, 24, 34, 103, 104, 107, 108, 110, 114, 119, 122, 128, 129, 132, 138, 149–52, 154, 158, 180, 201n4
pornography, 143
pragmatism, 170, 172
preference (homosexual), 3, 4, 7, 50, 51, 55–56, 60, 62, 67, 72–74, 75, 76, 77, 87, 93–95, 111–12, 130, 134, 158, 163, 198n3, 199n12
pregnancy, 38, 39, 63, 92, 113, 177, 203n13
prejudice, 178–79
prison, 8, 25, 27, 72, 146
Problemata, 1, 12, 17–19, 49, 114, 195n1, 196n1
projection, 152
pseudoscience, 175
psychiatric institutions, 27
psychiatric patients, 25–26, 40, 106, 141, 143, 152, 157, 165, 171
psychiatry, 12, 26, 31, 43, 44, 59, 136–38, 143, 145, 146, 151, 152, 155, 156, 171, 173, 174, 175
psychoanalysis, 26, 148–50, 153–55, 158, 202n3
psychology, 12; social psychology, 12, 14, 49, 185
Psychopathia sexualis (Krafft-Ebing), 13, 134, 143–46, 153

queer theory, 89, 175, 182, 195n2

race, 48, 179, 190, 202n8. *See also* racism
racism, 110, 155, 178–80, 182, 184
Rahman, Qazi, 46, 112, 127

rape, 90, 146, 170, 177, 202n2
Regnerus, Mark, 183, 186
regression, 142, 152
Renaissance, 110, 117, 201n5, 202n8
reproductive success, 13, 79, 84, 86, 99, 104, 106, 110, 111, 119–22, 163, 201n6
Richerson, Peter, 128–32
Rohy, Valerie, 183
Rome, 114, 201n5

sadism, 137, 146, 147, 153, 157–58, 202n2
Sambia people, 8, 87, 88, 115–17, 124
Samuels, Richard, 41
Samurai, 117, 124, 125, 126f
schizophrenia, 24–27, 106, 196n3
Schopenhauer, Arthur, 23
secularization, 176
seduction, 31, 175
sex, nonreproductive, 92, 138, 140
sexism, 179
sexology, 143, 148–51, 152, 202n4
sex segregation, 50, 62, 75. *See also* prison
sexual act/activity, 1, 28f, 44, 50, 58, 81–83, 85, 86, 115, 118, 122, 124, 147, 158, 159, 177, 199n18
sexual deviation, 113, 141, 142, 152, 153, 156, 157, 160, 163, 183
sexual identity. *See* identity (homosexual)
sexual instinct, 7, 137, 140, 141, 143, 146, 147, 149; as sixth sense, 141
sexual inversion, 6, 7, 57, 63, 103, 147
sexuality: extrinsic view of, 82; intrinsic view of, 82–83
sexual neuroses, 145
sexual orientation, 6, 8, 31, 34, 35, 37, 40, 62, 67, 68, 69, 74–78, 88, 92, 96, 109, 110, 123, 149, 155, 157–59, 165–67, 173, 175, 176, 180, 181, 183–87, 189, 191, 195n2, 196n5, 199n13, 204n5, 204n10
sexual perversion, 55, 137, 146. *See also* paraphilia
sexual pleasure, 17, 147, 148
sexual practices, 56, 87, 88, 150. *See also* sexual act/activity
sickle cell hypothesis, 99, 104–8, 111, 121, 123, 127
sin, 1, 3, 18, 118, 176

situational homosexuality, 8, 146
social construction, 42, 48, 49. *See also* constructivism
sodomy, 1, 6, 46, 115, 117, 145, 200n23, 201n1
Soviet Union, 173
Spanish flu, 24
Spitzer, Robert, 37, 40, 154–57, 202n9
Stein, Edward, 8, 12, 15, 70–72, 83, 96, 198n11, 200n22
sterilization, 24, 25
stigmatization, 6, 13, 166, 171, 175
sugar daddy hypothesis, 99, 108, 110, 111, 123, 127, 131

teleology, 33, 42
third sex, 22
Tours, Paul Moreau de, 141
transgender people, 113, 195n6
transgenerational homosexuality. *See* homosexuality: transgenerational or ritualized
Trumbach, Randolph, 117
twins, 24, 26, 27, 28f, 29, 29f, 30, 30f, 34, 35, 36, 38, 101, 196n2; dizygotic, 26, 27, 29, 30, 34–36; monozygotic, 26, 29, 34–36
typicality, 33, 42, 163

Ulrichs, Karl Heinrich, 20, 57, 137
unconscious internal conflict, 152
unisexual, 90
United States, 16, 40, 122, 150
unnatural, 13, 23, 53, 56, 57, 61, 82, 155, 162, 169, 175, 176
Urnings ("Uranists"), 57

VanderLaan, Doug, 110–11
Vasey, Paul, 72, 110
vegetarianism, 154, 155
Victorian era, 101

Waddington, Conrad, 37
Wakefield, Jerome, 163–69, 203n9
WEIRD (Western, Educated, Industrialized, Rich, and Democratic), 133, 134
Wilson, Glenn, 46, 112
World War I, 24, 150, 173
World War II, 25, 26, 152, 153

Yzerbyt, Vincent, 179

zoology, 3, 8, 10, 12, 51, 57, 60, 61, 81, 87, 88, 98, 137
zoophilia, 157. *See also* bestiality
Zuk, Marlene, 85, 89
Zulu people, 130